DATE DUE

Observing the Sun is one of the most interesting and rewarding facets of astronomy to which amateurs can contribute. Few areas of science offer as many opportunities to contribute meaningful data to the scientific community. It is the one branch of astronomy that requires only modest equipment and can be pursued during the day.

Peter Taylor is a keen and highly experienced observer of the Sun. In this book he explains in a clear and practical way everything that a telescope user needs to know in order to make solar observations. The author draws on his many years of personal experience as chairman of the Solar Division of the American Association of Variable Star Observers' American sunspot program.

The book deals with the following topics: historical background, choice of equipment for the safe conduct of solar observations, observations of sunspots and eclipses, and reporting observations. New techniques, such as the electronic recording of solar flares, are included. The level of presentation is understandable to anyone with basic astronomical knowledge and some experience in handling a small telescope.

Observing the Sun

The Practical Astronomy Handbooks are a new concept in publishing for amateur and leisure astronomy. These books are for active amateurs who want to get the very best out of their telescopes and who want to make productive observations and new discoveries. The emphasis is strongly practical: what equipment is needed, how to use it, what to observe, and how to record observations in a way that will be useful to others. Each title in the series will be devoted either to the techniques used for a particular class of object, for example observing the Moon or variable stars, or to the application of a technique, for example the use of a new detector, to amateur astronomy in general. The series will build into an indispensable library of practical information for all active observers.

Harold Hill   *A Portfolio of Lunar Drawings*
Kenneth Glyn Jones   *Messier's Nebulae and Star Clusters*, 2nd edition

Observing the Sun

PETER O. TAYLOR
Chairman, AAVSO Solar Division

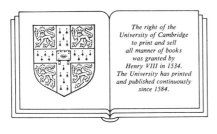

The right of the
University of Cambridge
to print and sell
all manner of books
was granted by
Henry VIII in 1534.
The University has printed
and published continuously
since 1584.

CAMBRIDGE UNIVERSITY PRESS
Cambridge
New York Port Chester
Melbourne Sydney

Published by the Press Syndicate of the University of Cambridge
The Pitt Building, Trumpington Street, Cambridge CB2 1RP
40 West 20th Street, New York, NY 10011–4211, USA
10 Stamford Road, Oakleigh, Victoria 3166, Australia

First published 1991

Printed in Great Britain at the University Press, Cambridge

British Library cataloguing in publication data available

Library of Congress Cataloguing-in-Publication Data

Taylor, Peter O.
    Observing the sun/Peter O. Taylor.
      p. cm. – (Practical astronomy handbook series; 3)
    Includes bibliographical references and index.
    ISBN 0-521-40110-0
    1. Sun—Amateurs' manuals.   I. Title.   II. Series.; 3
    QB521.4.T39   1991
    523.7—dc20   91-27809 CIP

ISBN 0 521 40110 0 hardback

Contents

Preface

George Ellery Hale, regarded by many as the father of modern solar astronomy, has said, 'No chapters in the history of science are more inspiring than those that recount the discoveries of the amateur. He works because he can not help it, impelled by a genuine love for his subject and inspired by an irresistible influence which he seeks neither to justify or explain. His reward lies in the work itself and in the hope that it may contribute to the advancement of knowledge ...' Thus, the world owes an immense debt to the amateur, and nowhere is this more apparent than in studies of the Sun.

Observing the Sun is one of the most interesting and scientifically productive endeavors in which an amateur astronomer can participate. Few areas in science offer amateurs the opportunity to contribute meaningful data to professional scientists as many solar observers do, and to have such a fine time doing it. Moreover, because of its nearness, and consequently its large angular diameter and abundance of light, detail on the Sun can be resolved with even a small (**but adequately protected**) telescope.

Many of the Sun's phenomena occur more or less regularly, but still vary enough so that they continue to keep our interest and anticipation at a high level. On the other hand, the Sun is frequently a source of surprise and excitement for even the most seasoned observer. The sudden and dramatic appearance of a great sunspot group which is many times the size of the Earth, or the sight of a huge eruption of magnetically charged gas stretching tens of thousands of kilometers into interplanetary space, are not soon forgotten.

Whatever the activity, both professional and amateur observers agree that the Sun often behaves in ways that are quite opposite of our expectations! As the English amateur astronomer, Richard Carrington, so aptly stated in his nineteenth century announcement of the changes in latitude of emerging sunspot groups during a cycle 'It will be found to be another instructive instance of the regular irregularity and irregular regularity which in the present state of our knowledge, appear to characterize the solar phenomena ...'

Even though our understanding of the Sun has increased greatly since Carrington's time, there is still a good bit of truth in his statement. As a result, the Sun continues to be a fascinating subject for each of us.

Amateur astronomers who wish to contribute the results of their observations to the scientific community cannot find a more rewarding object on which to base their investigations, and those who observe for enjoyment alone will not be disappointed. I hope that everyone who decides to participate in this intriguing area of astronomy will gain the knowledge and pleasure from this pursuit that I have, and that this book will become a useful addition to their library.

Peter O. Taylor*

*Present address: PO Box 5685, Athens, GA 30604, USA.

Introduction

During the years that I have been privileged to work with the contributors to the American Association of Variable Star Observers' Solar Division and American sunspot program, I have been asked many wide-ranging questions about the Sun. They have concerned sunspots and other of the Sun's phenomena, the program's work and background, and the history of solar observations in general. The inquiries have come from a variety of sources: newcomers and seasoned observers, members of the media, and those interested in simply learning about the Sun for their own enlightenment.

For better or worse, several courageous individuals have suggested that I prepare a single resource which would attempt to explain many of these topics in a way that is practical and informative for a serious amateur solar astronomer, but that would also aid the novice as he or she takes the first steps along a long and rewarding road. This book represents the result of my efforts to do that. (It hasn't been easy.)

The major portion of the book is divided into sections which deal with the following topics: (1) Appropriately referenced historical information and descriptions for many of the concepts that are fundamental to an amateur's study of the Sun. (2) Equipment and requirements for viewing the Sun safely, either directly or by projecting its image. (3) The concept of the relative sunspot number, details of the sunspot cycle and systems of sunspot group classification. (4) Techniques for determining the heliographic positions of sunspots and other solar features, and solutions for many common observing problems. (5) Procedures to gather and report potentially valuable data to the scientific community.

For those who are interested in observing the Sun electronically, a description of the construction and use of a very low frequency radio receiver which can be used to indirectly monitor the occurrence of solar flares in real time has been included, along with a similar description of a simple amateur magnetometer. Additional sections outline the techniques required to search visually for solar white-light flares, provide a brief overview of the major effects of the solar–terrestrial relationship, and discuss some practical aspects and projects for an amateur solar-eclipse chaser.

When a few simple and inexpensive precautions are taken the Sun can be viewed with the same degree of safety as other stars. But unlike other stars, these special procedures **must** be followed when observing the Sun. Consequently, safety has been of prime importance in those sections which deal with optical equipment and observational techniques. If I have erred in this area, it is on the side of protection. Chapter 6 is particularly important in this regard, and should be thoroughly digested **before** any observations of the Sun are attempted.

Those who are new to observations of the Sun will also be particularly interested in the material which is included in Chapter 7. A good deal of the information which is contained in these two sections is especially intended for use by novices, although it is hoped that more experienced viewers will also find portions of the material to be of interest to them.

In most cases, the book does not include sources for equipment or special accessories, although I do make specific recommendations whenever that is advisable. Information of this sort may be obtained from publications which deal with the popular aspects of observational astronomy, and through organizations which are regularly engaged in the compilation of information that relates directly to the observer's particular field of interest.

A substantial portion of the information on instrumentation and observational procedure was provided by two longtime contributors to the Solar Division's American sunspot program. They are Thomas A. Cragg, an astronomer at the Anglo-Australian Observatory in New South Wales and one of the original members of the Division, and David W. Rosebrugh, who until his death in 1988, made almost daily observations of the Sun for nearly forty-five years.

Thomas G. Compton, an experienced sunspot observer, and Dr Mary Jane Taylor, an astronomer at the Space Astronomy Laboratory, University of Wisconsin (and my daughter), took time from their busy schedules to read the manuscript and offer useful comments. Several of the superb photographs which are included in the book were taken at the National Solar Observatory at Sacramento Peak (SPO) or at other major observatories, and were supplied by Dr Donald F. Neidig of SPO, French collaborator Dr Jean Dragesco, Thomas Compton and others.

Arthur J. Stokes, who founded the Reuter–Stokes Company which designed and developed the X-ray soil sensors aboard the Viking Mars Landers, devised the solar flare monitor that is described in Chapter 12, and supplied the details of its construction. Well-known South African amateur astronomer, M. Daniel Overbeek, supplied information for his fine little magnetometer and recording device, and CompuServe® Graphics Manager Gregory W. Beach contributed the excellent schematic drawings of the electronics equipment.

I am indebted to each of these dedicated people for their interest, and for the advice and counsel which they offered to me during the preparation of this book. I would truly be remiss if I failed to also thank the authors of the many references from which I have borrowed freely. Without their obvious expertise and considerable knowledge the book simply couldn't have been written at all.

Last, but certainly never least, a round of applause is due to my wife Pamela for her extraordinary encouragement and support, and view from the writer's perspective. I'm sure that Pam has now learned more about the Sun (and about excited sunspot observers!) than she thought she would ever want to know.

It is the way of those who study the Sun.

1

Historical perspective

The history of the sociological and scientific relationship between man and the Sun is far too lengthy to be chronicled in a single chapter. Therefore, in the following paragraphs I will try to summarize those events which in my experience have generated the majority of questions pertaining to the history of solar observations and research. Since much of the interest in the past and present is centered on the study of sunspots and phenomena which are associated with them I have emphasized this field, although I have tried to present these aspects in context with other major discoveries about the Sun.

The earliest records show us that primitive cultures frequently thought of the Sun as an individual, or as a person carrying a huge ball of fire. Since the Sun appeared regularly and didn't seem to change, a number of fanciful tales arose in connection with its behavior, which to the ancients was more like that of a slave than that of a master. In his 1949 book about the Sun, the famous solar astronomer, Donald H. Menzel, relates one legendary explanation for this apparent contradiction: 'It seems that the Sun was once very erratic. Sometimes he hurried too fast on his journey; at other times he dawdled. On occasion he came too close to the Earth; often he was too far away. Sometimes he failed to appear at all. Finally, after great difficulties have been surmounted, the Sun is caught in a trap or net, beaten into submission, and thereafter performs his duties without remonstrance.'

Later civilizations believed that the Sun was a deity and pictured it in extraordinary ways, often as a god transported across the daytime sky within a ship or chariot (Noyes, 1982), or regarded it as proof of the validity of their philosophical beliefs. Nowhere is this more evident than in the glorification of the Sun so loudly proclaimed by the powerful philosopher Aristotle, whose insistence upon its perfection profoundly influenced the thought and direction of early western society.

Because the Sun plays such an enormous role in our lives, it has always been viewed with fascination and curiosity. To its earliest observers, the Sun generally appeared to be nearly the same each day; however every once in a while a dark smudge would appear on its face which would cause great consternation before departing as swiftly as it came. According to Schove (1983), the earliest of all references to these large spots is contained in a translation of a Chinese oracle bone from the twelfth century BC: 'Will the Sun have marks? It really has marks ...'

Ironically, the first written record of a sunspot was made by a pupil of Aristotle named Theophrastus, who observed a spot on the Sun in the fourth century BC (Noyes, 1982). The Chinese, and pre-Historic Peruvian astronomers began to make and record their observations of the spots and other solar phenomena within a few centuries of Theophrastus's sighting, but for a variety of reasons the attempt to record them from Europe came much later. For that matter, even the historical observations made by the Chinese were not available in a European language until after 1873 (Giovanelli, 1984).

Today, many historians believe that the long delay in western reports can be traced to the teachings of Aristotle and his claim that the Sun was literally a ball of pure fire which could not be blemished. Despite the identification of sunspots as features on the Sun's surface by Aristarcus in the fifth century AD (Noyes, 1982), the strength of Aristotle's instruction combined with the unwillingness of academics to refute it, hid the truth for over a millennium. This unfortunate situation was not fully resolved until sometime after the first telescopic observations of the Sun.

The first observations of the Sun's spots were made with the unprotected eye during the morning or evening and when the Sun was obscured by smoke or haze so that its brilliance was somewhat diminished. (**The reader is cautioned against observing the Sun without adequate eye protection!**) The large sunspot groups which can be seen in this manner (Figure 1.1) typically occur

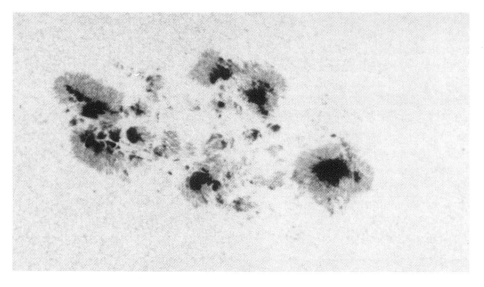

Figure 1.1 This large sunspot group grew to encompass an area of over 5.5 thousand million square kilometers during May 1990, making it easily observable with the unaided (but suitably protected) eye. Photograph courtesy of Jean Dragesco.

within a year or so of the sunspot cycle maximum. This aspect has proven to be a valuable one for those who study the spot cycle since it has allowed astronomers to determine the approximate dates of cycle maxima and minima which occurred a century or more before sunspots were first viewed telescopically.

Prior to the time that their physical connection with the Sun was universally acknowledged, observations of these 'naked-eye' sunspots were usually attributed to transits of known or undiscovered interior planets, or to phenomena which occurred in the Sun's or Earth's atmosphere. (A *transit* occurs on those occasions when an object passes directly between the Sun and Earth.) Even Galileo's friend, the famous mathematician and astrologer Johannes Kepler, incorrectly attributed a spot which he viewed in 1609 to a transit of Mercury, although he eventually acknowledged his error, allowing that he was 'mistaken.'

Although the spots were recorded in this simple manner for over two-thousand years, it was not until 1610 and the advent of the astronomical tele-scope that the real character of these secular features became apparent. At that time, Galileo Galilei, surely one of the greatest intuitive scientists that the world has ever known, seized upon and perfected the newly invented telescope and began to make routine observations of the Sun and other astronomical objects.

Shortly thereafter use of the device spread to several of Galileo's con-temporaries, among them Goldsmid (Fabricus), Scheiner and Harriot. Their investigations, coupled with those by Galileo, soon began to unlock the secrets of the Sun. Unfortunately, because of social and political pressures brought about by the jealousy of his fellow astronomers and by his own combative nature, Galileo felt compelled to delay the announcement of his studies. As a result Fabricus became the first to publish findings which showed that sunspots were physically associated with the Sun (Bray and Loughhead, 1965). Interest-ingly, Fabricus observed the spots by projecting the Sun's image into a darkened room with a camera obscura, or pinhole camera, instead of viewing them directly.

Perhaps the real reason behind the reluctance of the powerful leaders of the day to acknowledge the true nature of these features had more to do with granting a scientist the right to teach and publish their views than with an affront to the supposed perfection of the Universe (Drake, 1957). But whatever the cause, the effect was the same: to temporarily retard progress in the area of scientific discovery and inhibit the flow of knowledge in the western world.

Even though Galileo delayed the dissemination of his conclusions regarding the nature of the spots, he is rightfully credited with an amazing degree of perception concerning many of their characteristics. He was almost certainly the first to interpret the Sun's spots as solar, rather than atmospheric or planetary phenomena, and realized early on that they almost always occurred in circum-solar bands which extend thirty degrees or so from the equator.

Furthermore, by 1612 Galileo had tracked the spots and the bright cloud-like

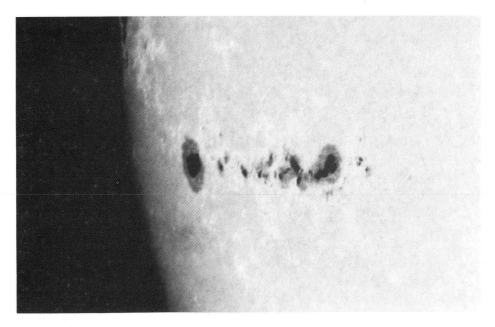

Figure 1.2 The light-appearing active features of the Sun called faculae are more easily seen in association with sunspot groups near the Sun's limbs, such as demonstrated in this fine photograph taken by Jean Dragesco.

features of the Sun called 'faculae' (Figure 1.2) across its disk, and by doing so had demonstrated that the Sun rotated on its axis. Moreover, he had skillfully argued these points (and effectively demolished his critics while doing so) in his replies to Scheiner, who believed that the spots were objects located between the Sun and Earth. Galileo accomplished this in a way which was typical of him; through a series of letters written to the wealthy merchant and amateur scientist Mark Welser, in response to those written anonymously by Scheiner (Drake, 1957). (Any reader who is interested in the enormous impact of Galileo upon society and the scientific community is urged to obtain Stillman Drake's fine discourse on this amazing individual.)

Most of Galileo's work on the Sun took place in the few years which followed the introduction and initial development of the telescope. On the other hand, Scheiner continued his observations for many years and eventually compiled the information, along with a number of fine engravings which showed the spots and faculae as they traversed the Sun's disk, into his great work, *Rosa Ursina sive Sol* (1630).

While Scheiner does demonstrate an understanding of the tilt of the Sun's axis in the publication, it is likely that Galileo was aware of the orbital relationship between the Sun and Earth many years beforehand. In the *Dialogue*, Galileo uses this argument to explain the movement of the Earth according to circumstances which could not have taken place after 1613 (Drake, 1957). In light of this,

4

Scheiner's most important contribution to the study of the Sun is likely to have been the observation of the spots themselves, since these data have been used to help define the spot cycle; particularly the maximum which occurred in 1626.

Unfortunately, after the flood of new information which arose from the use of the telescope between 1610 and 1630, the Sun itself intervened in the discovery process with a long lull in activity that lasted into the following century. Hardly any spots were seen during this period which extended from 1645 until 1715, and eventually came to be known as the 'Maunder Minimum' after the English astronomer who investigated it in detail.

Then in 1769 a University of Glasgow professor and astronomer named Alexander O. Wilson completed what is regarded as the first scientific investigation into the *properties* of sunspots, rather than simply registering their number or location (Bray and Loughhead, 1965). Wilson found that the appearance of large spots near the Sun's limb was saucer-like, and explained his findings as depressions in the Sun's surface which were caused by a lack of material covering a dark and solid inner-core. In 1770 Wilson won the astronomical award given by the Science Society of the University of Copenhagen for this project. (The effect is now thought to be caused by the higher transparency of the spot material relative to the Sun's 'surface,' or photosphere.)

At the time that Wilson offered his findings, he believed that the Sun was surrounded by two atmospheres; an inner one blanketing a presumably cooler planet-like object where life might well exist (!), and a hot outer shell (Chambers, 1890). The outer atmosphere was luminous, and consequently it came to be called the *photosphere* (the luminous, or light-sphere). The photosphere is now known to be the visible exterior of the Sun; a region which is several hundred kilometers thick with a surface temperature that is approximately 5800 kelvin.[1]

Of course the photosphere is not the smooth glowing shell which Wilson and other early scientists envisioned. Instead it is composed of many bright areas, each of which is several hundred kilometers in diameter and is surrounded by darker lanes of cooler material (Giovanelli, 1984). These irregularly shaped features are known as the solar 'granules' (Figure 1.3), a term which is thought to have originated with Dawes in the nineteenth century who described the effect as 'irregular luminous clouds surrounded by less brilliant lanes ...' (Todd, 1899). A similar view was expressed in 1792 by the Danish astronomer, Thomas Bugge, who may have actually been the first to observe the effect, and also by William Herschel at about the same time (Webb, 1893).

The granules actually form the tops of huge currents which rise from beneath the Sun's exterior at a rate somewhere near 500 meters per second (Zirin, 1988) and erupt at the surface in a manner which is similar to a thick bubbling liquid. Typically they are short-lived; most have lifetimes which are measured in

---

[1] The kelvin scale is derived by adding the centigrade temperature to 273.15. Thus 100 degrees centigrade is equivalent to 373.15 kelvin.

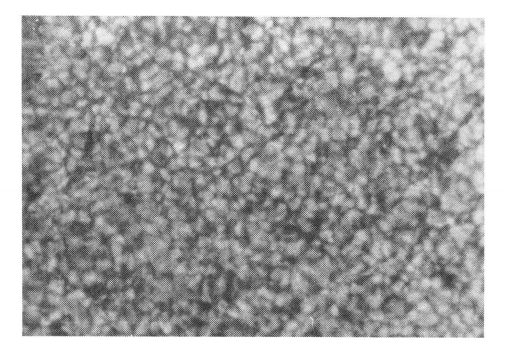

Figure 1.3 Most of the Sun's surface, or photosphere, is made up of a network of cells known as the solar granulation. Note the dark intergranular lanes which surround each of the cells. Photograph courtesy of Thomas G. Compton.

minutes before they cool and decay. When seen with lower magnifications, the granules produce a visual effect known as the 'rice-grain' pattern, a description which was first suggested by Huggins in his summary of the phenomenon at the Royal Astronomical Society meeting of 1866 (Huggins, 1866).

The Danish astronomer and director of the famous Round Tower Observatory, Christian Horrebow (1718–1776), was probably the first to suspect a periodicity in the numbers of sunspots (Vitinskii, 1965). In the last year of his life, Horrebow entered the following remarks into the observatory protocol, indicating that he believed a more systematic observation of sunspots might lead to 'the discovery of a period, as in the motions of the other heavenly bodies ...' and added, 'then, and not till then, it will be time to inquire in what manner the bodies which are ruled and illuminated by the Sun are influenced by the sunspots ...' (Young, 1888).

The study of sunspots was a subject of recurring interest at the Round Tower Observatory from the time that it was founded in 1637 on the recommendation of the astronomer and only pupil of Tycho Brahe, Longomontanus, until its eventual demise as a scientific institution. Peter the Great viewed sunspots from the Tower on several occasions under the guidance of Christian's father, Peder Horrebow, during his tenure as observatory director. News reports of the day indicate that the Czar actually rode the observatory's long wooden ramp to the top of the tower on horseback, accompanied by the Czarina who was driven up in a carriage drawn by four horses! (Thykier, 1988).

Unfortunately, the majority of Christian Horrebow's work was lost for a considerable period of time in the destruction of Copenhagen during the Napoleonic Wars. As a result, it was not until the mid-nineteenth century that the existence of a *sunspot cycle* was recognized. At that time the German amateur astronomer and apothecary, Heinrich Schwabe, published his long series of sunspot observations and suggested that their number appeared to vary regularly with a period of about ten years (Schwabe, 1849, 1851). His important discovery forms the basis for an often-told tale of scientific serendipity.

It seems that Schwabe's real purpose was to discover an additional planet orbiting the Sun within the path of Mercury. His forty-three-year record of sunspot observations was made in support of that research, so that he could differentiate between sunspots and the expected transit of his proposed planet. Although he never found his missing world, Schwabe's dedication and perception guaranteed him a place in astronomical history and caused him to react to his unexpected discovery by stating, 'I can compare myself to Saul, who went out to find his father's asses and found a throne ...' (Noyes, 1982).

Soon after Schwabe's announcement, the director of Bern Observatory, Rudolf Wolf, became interested in Schwabe's work and also began to study the spots. Wolf would have preferred to compute the spot group's areas rather than count the number of spots, but his equipment was not suited to this purpose. Consequently, Wolf developed an arbitrary counting system and called the resulting index 'Universal' sunspot numbers (Wolf, 1858). Eventually these came to be known as relative sunspot numbers, which today form the longest continuous record of the Sun's activity.

Through an extensive search of the early records, Wolf was able to derive a more accurate value of a little over eleven years for the average length of a cycle (Wolf, 1852). It is interesting to note that according to Schove (1983) Chinese observations of naked-eye sunspots between AD 188 and 1638 also show a period of around eleven years. During Wolf's exploration of the old observatory records, Horrebow's unpublished diaries were finally recovered, and the material was published shortly thereafter (Thiele, 1859). Horrebow's notes clearly showed the cycle minima which transpired in 1766 and 1775, and also indicated that a maximum had taken place around 1770 (the maximum actually occurred during the latter half of 1769).

In 1855 Wolf became director of the Swiss Federal Observatory in Zurich, Switzerland, where he and his successors, Wolfer, Brunner and Waldmeier, continued an expanded program of sunspot observation and research. During his term as director of the observatory, Wolf also instigated an international collaboration of observers whose data were employed when the Swiss weather was poor. However, the long series of values known as *Zurich Relative Sunspot*

Numbers are mainly the result of observations which were secured at the observatory in Zurich or by its official sub-stations at Locarno and Arosa, all of which employed identical instrumentation.

Almost a decade after Schwabe published his discovery, an English amateur astronomer named Richard Carrington presented the results of his research on the positions of the spots. Carrington, the son of a brewer, had originally intended to become a minister but became interested in astronomy while attending Trinity College in Cambridge. His interest in the Sun was sparked by Schwabe's identification of a period in the numbers of sunspots, and continued until he was forced to abandon observations and assume control of the family brewery in 1861 on the death of his father (Bray and Loughhead, 1965).

Carrington's observations showed that a relationship existed between the heliographic (solar) latitude of emerging sunspots and the phase of the newly discovered spot cycle (Carrington, 1858). This phenomenon, in which new spots erupt at high sunspot latitudes at the beginning of a cycle and then at progressively lower locations as the cycle matures, has come to be regarded as one of the principal characteristics of the sunspot cycle. During the time that Carrington compiled these observations, he also became one of the first to detect the Sun's differential rotation (Webb, 1893), which results from the Sun's gaseous make-up.

In the second investigation, Carrington found that the Sun rotates more slowly at polar latitudes where a single revolution requires about a month to complete, than it does at the equator where the rotational rate is some five days less. It is interesting to note that, according to Webb, Carrington determined the existence of both of these effects by using only a simple cross-wire within his instrument; no micrometer or other special measuring equipment was employed.

Eventually Carrington's findings concerning the emergence of spots during a cycle were corroborated by Gustav Spörer, and thoroughly investigated by Edward W. Maunder who demonstrated the existence of the latitude-effect beyond all doubt (Maunder, 1904). Consequently, the phenomenon is generally called *Spörer's Law*, while the graph which results when sunspot locations are plotted according to time is known as a *Maunder butterfly-diagram*. The original butterfly-diagram was drawn by Maunder and his wife Annie (Maunder, 1940), just prior to the Royal Astronomical Society meeting in 1904 where Maunder presented his famous paper, and is currently displayed in the library of the US National Center for Atmospheric Research.

It is intriguing (and not a little unusual!) that Carrington was also the first to observe and record a solar flare (Carrington, 1859). These rare, but spectacular events are known as solar white-light flares when they are seen in the normal visual portion of the spectrum, as this flare was. Fellow Englishman Richard Hodgson independently observed the same event which was followed by an intense geomagnetic storm and brilliant aurorae. Although Wolf had previously

Figure 1.4 Near the Sun's limbs, the huge eruptions of solar gasses called prominences can be viewed in the red light of atomic hydrogen. The small filled circle indicates the relative size of the Earth in this photograph by Jean Dragesco.

uncovered a correlation between the sunspot cycle and geomagnetic disturbances, the dramatic effect of these combined occurrences formed the impetus for future investigations into the true nature of the solar–terrestrial relationship (Noyes, 1982).

At the 1868 eclipse of the Sun in India, French astronomer P. J. Janssen made another important discovery. During the eclipse, Janssen had observed a huge eruption of gas on the Sun's limb, a feature known as a solar prominence (Figure 1.4). He had viewed the phenomenon spectroscopically, carefully noting the presence of bright spectral lines near the wavelength of atomic hydrogen in the red portion of the spectrum.

After the eclipse was over Janssen again pointed his instrument towards the location of the prominence, and again he found a bright area. After the spectroscope's slit was opened slightly, the shape of the prominence could clearly be seen. The identical discovery was made independently and virtually simultaneously by the English astronomer Norman Lockyer (who did not attend the eclipse), and ultimately led to the invention of the spectrohelioscope. After this instrument was developed, details on the Sun's disk could be discerned for the first time, and Janssen was rewarded with the directorship of France's new Meudon Astronomical Observatory.

While he was director of the Meudon Observatory, Janssen obtained a number of extraordinary photographs of the solar granulation. (No photographs of the Sun of any type were made until after 1870.) A few of the photographs appeared to show even the finest granular structure, but eventually this was found to be a consequence of distortions produced in the Earth's atmosphere. As a result, the first high-resolution photographs of the granular network were delayed until the balloon experiments which were conducted by Schwarzschild (1960).

At about the time that Janssen was photographing the solar surface, American physicist Jonathan Lane developed an idea which showed that the Sun was not the planet-like object which Herschel and Wilson had envisioned. Instead, Lane suggested that the Sun was entirely composed of gas held together by the force of its own gravity, and driven by an internal source of energy. The origin of this dynamic mechanism was originally characterized as nuclear in nature by Arthur Eddington in 1926, but a complete definition of the process was delayed until after the end of World War II (Giovanelli, 1984).

George Ellery Hale, the guiding force behind the construction of the Yerkes, Mount Wilson and Palomar Observatories, and a developer of the spectrohelioscope, discovered the magnetic nature of sunspots in 1912 (Hale, 1912). Hale accomplished this by noting that the spectral lines which arise from the spots are split into two or three separate lines. The same spectral line-splitting and polarization in sunspot spectra is also seen in laboratory spectra when their source is exposed to a strong magnetic field. Hale correctly deduced that the presence of this condition (known as the *Zeeman-effect* after the Dutch physicist who discovered it in 1896) in sunspot spectra indicated that the spots are magnetic phenomena.

By 1919, Hale and his co-workers at Mount Wilson had gone on to describe the polarities of individual sunspots and their reversal from cycle to cycle (an effect which is termed the *Hale–Nicholson Law*), and had established the Mount Wilson magnetic classifications (Hale *et al.*, 1919). In spite of its age the Mount Wilson system continues to be used by professional astronomers, although because special equipment is required to obtain these data, it is not suitable for use by amateur astronomers.

In 1938 M. Waldmeier devised the evolutionary sunspot-group classification scheme for visual observers which continues to be in widespread use today (Waldmeier, 1947). This technique for grouping sunspots into clusters according to their physical appearance and size is generally referred to as the 'Zurich Classification' system. The categories which are assigned to spot groups according to this method have proven to be especially valuable for studies of the growth and decay of spot clusters as well as for research into the physical characteristics of sunspots.

In the early 1950s, Ludwig Biermann (1951) suggested that there was evidence that particles streaming away from the Sun, rather than the pressure of sunlight, were the reason that the tails of comets always pointed away from the Sun. Later in the decade, the American physicist Eugene Parker showed that this was indeed the case, and dubbed the effect the 'solar wind' (Noyes, 1982). A few years later the presence and strength of the wind was recorded by American and Soviet satellites, and conclusively demonstrated by experiments aboard the Mariner 2 spacecraft on its journey to Venus.

However, the Mariner project also showed that the wind's speed varied greatly, and frequently occurred in 'gusts.' Eventually, just before and during the Skylab mission, the cause of the effect was traced to high-speed particle streams which originate in large-scale, very low density active areas which occur within the Sun's atmosphere, or corona. These openings in the atmosphere, referred to as coronal holes, are now thought to be the cause of periodic geomagnetic disturbances, and according to Hewish, Tappin and Gapper (1985) may well be the *only* mechanism to produce these storms. Thus, the solar wind and the phenomena which cause it to vary have come to play a fundamental role in our understanding of the solar–terrestrial relationship.

During his stewardship of the Swiss Federal Observatory, Professor Waldmeier completed the exhaustive search of the historical records begun by Wolf, and extended the sunspot record through 1960 (Waldmeier, 1961). More recently, John McKinnon of the National Oceanic and Atmospheric Administration has updated Waldmeier's tables and original reference to include data through 1988 (McKinnon, 1987, 1988). A plot of monthly-mean sunspot numbers between 1749 and 1988 is shown in Figure 1.5. Since auroral activity in middle and low latitudes can be correlated with the sunspot cycle in a manner which is similar to the occurrence of naked-eye sunspots, D. J. Schove has been able to project the dates of cycle maxima and minima back to AD 1500 (Schove, 1979, 1983).

Furthermore, recent research by Eddy, Williams and others, appears to offer indirect evidence of the existence of a sunspot cycle into the very distant past. For instance, in the 1970s John Eddy found that deviations in the amount of atmospheric carbon-14 measured in tree-rings thousands of years old, could be used to infer the level of the Sun's activity long before it was first monitored telescopically (Eddy, 1976). While individual spot maxima and minima cannot be pin-pointed by this method, the trend of changes in ^{14}C over the past 2000 years is supported by evidence which has been obtained from empirical techniques.

During the decade which followed Eddy's research, an even longer record of the Sun's activity was uncovered. This evidence appears in the study of ancient 'varves', annual layers of sediment which have been deposited by water run-off over many millions of years. George Williams (1985) and other investigators have determined that variations in the thickness and number of layers could

11

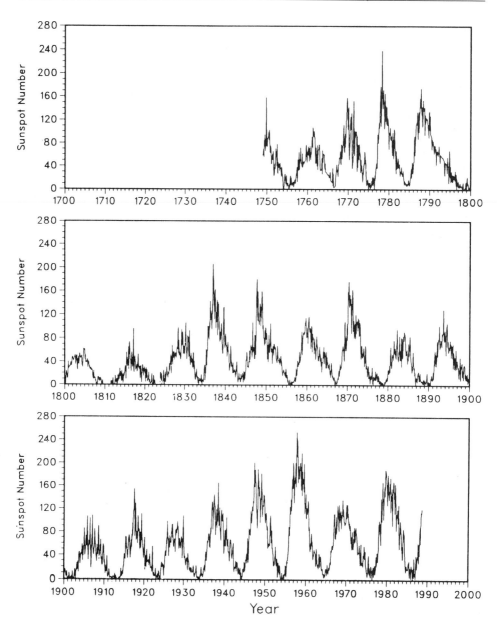

Figure 1.5 Monthly-mean sunspot numbers between 1749 and 1988 clearly outline the eleven-year spot cycle as shown in this diagram supplied by John McKinnon of the World Data Center A for Solar–Terrestrial Physics.

result from changes in the Sun's output. If this is correct, the record indicates that sunspot cycle activity has remained substantially the same for nearly 700 million years.

After his earlier invention of the *magnetograph*, an instrument which enables

12

astronomers to determine the polarities of magnetic concentrations which exist across the Sun's surface, H. W. Babcock proposed a model for many of the solar magnetic phenomena which had been observed up to 1960 (Babcock, 1961). Among other things, the model that Babcock developed suggests that the reversal of polarity which accompanies each new cycle results from the poleward drift of the magnetic concentrations which were originally contained within the *following* (most easterly) sunspot in a bipolar spot group.

Although some of the observed spot characteristics are not fully accounted for by the theory (Giovanelli, 1984), the Babcock model with revisions and additions by Leighton (1964, 1969) has been widely accepted until recently when it has lost some adherents. As a result, several different explanations for the Sun's magnetic phenomena have been proposed. One of the more interesting is the approach of Snodgrass and Wilson (1987), who suggest that sunspot and other of the Sun's magnetic features result from the actions of huge convective rolls which originate at the poles and migrate to the equator over a period of eighteen to twenty-two years.

Presumably, by the time that the rolls have reached the middle and lower latitudes they have compressed the magnetic fields (which rise from deep within the Sun's interior) sufficiently so that they manifest themselves in the eruption of sunspots, and the spot cycle begins. According to Snodgrass and Wilson, the rolls may also have an effect upon the Sun's atmosphere which could affect the rate of geomagnetic disturbances.

Since the retirement of Professor Waldmeier the work of the Zurich sunspot program has been carried on at the Royal Observatory of Belgium. Values derived at the new location are called *International*, rather than Zurich, Relative Sunspot Numbers. They are still determined in the manner which was originally devised by Wolf, although the technique was mildly modified in 1882 by Wolf's successor, Wolfer, so that it also accounted for the spot's structure.

The history of the *American Relative Sunspot Number*, while of considerably shorter duration than the long effort at Zurich, is a colorful one. The American program was initiated in 1944 as a wartime measure (the effects of solar activity on radio communications and in other critical defense areas were well-known by this time) and to provide a second, and fully independent sunspot index for the American scientific community. Two of astronomy's fabled members, Harvard Observatory's Harlow Shapley and Donald H. Menzel, proposed and supported its establishment; and J. Virginia Lincoln, Alan Shapley and others at the National Bureau of Standards developed the statistical techniques which are employed in the reduction of American Relative Sunspot Numbers.

The program's original observers were recruited from the ranks of the American Association of Variable Star Observers, whose headquarters was located within Harvard College Observatory. The program was administered by the Association through its *Solar Division*, although it was directed by Neil J. Heines, who received a salary from the United States Government (D. W. Rosebrugh,

13

Table 1.1

Radius: $R_\odot = 696\,000$ kilometers (109 Earth radii)
Gravity at the surface: $g = 2.74 \times 10^4$ centimeters/second/second
Total radiation: $L = 3.86 \times 10^{33}$ erg/second
Distance from the Earth: 1.496×10^{13} centimeters (214.94 $R_\odot$ or 1 AU)
Temperature of photosphere : ~5800 kelvin
Temperature of typical sunspot: ~4800 kelvin
Temperature of (Hot Thermal) solar flare: $T = 3\text{--}4 \times 10^7$ kelvin

personal communication). Subsequently, the Division has been headed by Harry Bondy, Casper Hossfield and (briefly) Robert Ammons. I have guided the effort since 1980.

The American program has operated continuously since its inception, and in the process has become international in scope. Perhaps most importantly, the analyses of American Relative Sunspot Numbers and solar flare activity now benefit greatly through the regular contributions of a vastly increased network of observers who are located on six (and occasionally seven) continents, so that the Sun is monitored over an entire day.

The long relationship between the program and the National Oceanic and Atmospheric Administration (NOAA) has continued. The results of the program's investigations are regularly supplied to NOAA (who support these efforts through an annual grant to the Association), and to many scientific organizations and interested individuals throughout the world.

Current research on the Sun is dominated by sophisticated techniques which measure the Sun's entire electromagnetic spectrum, from the shorter X-rays to the longest radio-waves, as well as the intricacies of its magnetic and atmospheric phenomena. Today, satellites, spacecraft and experiments carried by the Space Shuttle combine with other highly technological procedures to monitor the Sun's activities continuously. However, each of these endeavors builds upon the accomplishments of the pioneers in solar research, whether they be the remarkable intuitive abilities of Galileo, the discoveries of Carrington, the perception of Lane or the countless skilled observations obtained by both amateur and professional astronomers. While much has been learned in the intervening years, the challenge of the future remains the same as that of the past: to truly understand the Sun, and by so doing, the world around us.

Several of the more important features of the Sun and its phenomena are summarized in Table 1.1, which has been compiled from material presented in Zirin (1988).

2

The relative sunspot number

In the mid-nineteenth century, a German amateur astronomer named Heinrich Schwabe announced that his long series of observations showed evidence that the number of sunspots peaked at ten-year intervals (Schwabe, 1849). Intrigued by the possibilities of a periodicity in the numbers of sunspots, the director of the Bern Observatory, Rudolf Wolf, began to observe the spots himself. Shortly thereafter, Wolf initiated a systematic program meant to record each day's sunspot activity. The concept of the relative sunspot number, which Wolf originated and called *Universal Sunspot Numbers*, evolved directly from these early interests.

According to Bray and Loughhead (1965), Wolf would actually have preferred an index which was based on the accumulated area of the sunspot clusters, since he believed it to be the more accurate indicator of the Sun's activity. However, his limited equipment prevented him from drawing the projected image of the Sun to the correct scale, so Wolf necessarily abandoned the attempt to record the areas of spot groups.

At the same time, he realized that counting only the number of groups would fail to allow for the large diversity in size and structure among individual spot complexes, and would not produce a useful index. Similarly, Wolf believed that counting only the number of spots would also be unsatisfactory since large numbers of spots within an existing group would dilute the importance of spots in newly emerging regions. In light of these difficulties, he devised a simple process whereby these two components could be combined. The result is a sunspot number index that can easily be determined with a moderately sized telescope, and which also provides a reliable indication of the Sun's overall level of activity.

Since Wolf worked alone for the most part, he chose to count only the spots which he could see without too much regard for the variations in local weather conditions. Today, it is generally believed (e.g., McKinnon, 1987) that Wolf counted all of the groups with penumbra, and most of the larger bipolar clusters with simpler structures. The fainter groups, which could be seen only during those times when the observing conditions were excellent, were generally ignored.

Wolf's interest in the sunspot phenomenon continued after he assumed the directorship of the Swiss Federal Observatory in Zurich in 1855, and throughout his career. After his retirement the program was carried forward by his successors until December 1980, and the retirement of M. Waldmeier, who directed the observatory after World War II. From that time forward, the responsibility for determining the daily sunspot number has been assigned to the Royal Observatory of Belgium where it operates under the guidance of Dr Andre Koeckelenbergh.

Towards the end of the nineteenth century, Wolfer, who had followed Wolf as director of the Swiss Federal Observatory, altered the original method so that the smallest spots visible under excellent seeing conditions were included in the counting process. Wolfer also began to weight the more complicated spot groups according to their size and structure (McKinnon, 1987). As a result, present-day relative sunspot numbers are based on a count of all spots within the limit of resolution of a small telescope. This stipulation usually includes spots which equal or exceed about three arc-seconds in diameter as seen from the Earth. (One arc-second near the center of the Sun's disk represents a distance of about 700 kilometers.)

Today, Wolf's program and its newer counterpart, the Division's American sunspot program, utilize the efforts of many skilled contributors from locations throughout the world. In fact, an international collaboration of this sort was initiated by Wolf himself in an effort to eliminate gaps in the record due to the poor Swiss weather conditions. The International Astronomical Union has designated the indices of solar activity that result from these programs as R_1 for the continuation of the old Zurich series (now called International Relative Sunspot Numbers), and R_a, for American Relative Sunspot Numbers.

The American and International programs compute their values in a similar manner, although the American system does not weight individual clusters as we have outlined above. In this sense, the American values are actually 'Wolf numbers,' although both programs have adopted the first of Wolfer's changes, and include the fainter groups in the counting process.

The counting procedure is a simple one. When compiling the daily estimate, an observer first defines each isolated spot cluster as an individual group according to specific guidelines. For many years the grouping scheme for visual sunspot observers has been the evolutionary classification system outlined by Professor Waldmeier in 1938 (Waldmeier, 1947), which is described in a later chapter.

A group may contain just one, or well over a hundred individual spots. The clusters vary in size from a single spot which is only a few arc-seconds in diameter, to huge spot-complexes with east to west spreads that extend over thirty or more degrees of solar longitude and occupy thousands of millions of square kilometers of the Sun's 'surface.' In spite of these great differences among groups, Wolf's method requires that each be awarded a group value of *ten*.

After a cluster has been defined, the number of distinct spots within its borders are counted and the total is added to the group value. In this way, a single, small isolated spot is assigned a total value of eleven (the group value of ten, plus one spot). A large spot-complex is treated in exactly the same manner: its group value of ten is added to the number of individual spots within the group. Values for all sunspot-complexes on the visible hemisphere are similarly determined, and the grand total is recorded as the observer's estimated sunspot number, R, for that day. This simple empirical relationship may be stated as

$$R = 10g + s,$$

where g represents the number of groups and s is the total number of spots.

These data are then 'scaled' through the application of a statistical parameter, K. Each contributing observatory is assigned a specific K-factor based on the relationship of their data to a standard. The standard which is chosen depends upon the program that compiles the observer's data, but more often than not it is made up of a like portion of American or International relative sunspot numbers. This factor, or observatory constant, is subsequently used to correct each estimate so that it is as comparable as possible with the index for which it will be used, *and* the original Wolf series. The method of computation of observer statistical factors for the American program is explained in Chapter 13.

Furthermore, when the original technique of counting sunspots was refined by Wolfer in 1882, it became necessary to scale each count thereafter by a value of 0.60 in order to insure that it is homogenous with the earlier series. Thus K is assumed to be 1.00 for the counts made by Wolf, and *all* values after that time have undergone the reduction which makes them compatible with his determinations.

In addition to these requirements, K is also intended to compensate for an observer's individual equipment, judgement and local weather conditions. Even a small difference in instrumentation can produce a rather large disparity in the numbers of spots which are seen. In order to maintain the statistical integrity of this long series of values, Wolf and his successors employed a refractor of eighty millimeters aperture and strongly encouraged their supporting stations to do likewise. Today, however, a large diversity of equipment is employed by the members of the international observer networks, and it is all the more important that the spot counts be equalized through the use of rigidly defined observatory constants.

Personal judgement and average local observing conditions also differ among observers, and each can play a large role in the counting process. Even when observers utilize the same equipment under equivalent conditions, estimates are likely to differ. For instance, one observer will see only a few spots (or none at all), while another will count many; and the two may separate an active spot area into groups in different ways. These discrepancies must also be minimized through use of the K-statistic.

Reliable daily relative sunspot numbers have been determined continuously since 1848. Since sunspot observations from the time of Galileo and Fabricus in the early seventeenth century until 1848 were reconstructed by Wolf, they are considered to be less reliable than later values. These data also suffer in that they reflect the large seasonal changes in weather of Central Europe where most of the observations were made, rather than actual variations in the numbers of spots on the Sun.

In spite of these drawbacks however, the data are sufficiently accurate so that the maxima and minima of sunspot cycles which occurred between 1610 and 1848 can be established. Cycle parameters which pre-date telescopic observations of the Sun have generally been inferred through correlations between the sunspot cycle and the amount of naked-eye sunspots or mid-latitude aurorae (e.g., Schove, 1983), since these indices also peak around the time of sunspot maximum and mimic the rise and fall of each cycle in a general way.

The importance of the relative sunspot number lies in its simplicity and consistency, and in studies of its regular (and irregular) variations. During the long interval that it has been determined, Wolf's sunspot index has been found to accurately describe the level of solar activity for the entire visible disk of the Sun. Therefore today's index has become more valuable for predictive purposes than the original Wolf system of counting only the larger spots, or of the sunspot-area index which is subject to variations in measurement technique and is far more difficult to calculate.

While the relative sunspot number is an arbitrary concept, it was fortunately chosen. The present system of determining these data yields a statistic which is also highly correlated with geophysical activities. A growing number of terrestrial occurrences ranging from the effects on astronauts and spacecraft to those on communications and possibly the weather, are closely linked to various solar phenomena. These events in turn, are directly related to sunspot activity as it is expressed by Wolf's original concept.

3

Characteristics of the sunspot cycle

The sunspot cycle is one of the most fascinating of the Sun's features, and its study is all the more intriguing because the intensity of a spot cycle provides a good indication of the level of activity for many of the Sun's other active phenomena. For these reasons, a knowledge of the cycle is of fundamental importance to our understanding of the Sun, and of the terrestrial effects which result from its emissions. Recording the growth and decay of each cycle is an especially attractive project for the amateur solar astronomer since it may be monitored with very modest equipment, and the data which are acquired in this way are useful to researchers in a number of fields of study.

Each sunspot cycle is numbered according to a scheme which was devised at the Swiss Federal Observatory in Zurich, Switzerland. According to this system, the epoch of minimum for the first numbered cycle occurred in 1755 (Schove, 1983). The information which is necessary to define the physical characteristics of each cycle is usually determined from an analysis of relative sunspot numbers or by other indices which measure its growth and decay at regular intervals.

A cycle's activity level can also be measured by recording the fluctuation of the Sun's radio output at a frequency of 2800 megahertz (10.7 centimeters) and at other wavelengths. Solar radio flux emissions originate in the higher regions of the chromosphere and lower corona and generally parallel the sunspot index (Figure 3.1). However, the observed flux values vary by as much as (±) seven percent because of the continuously changing distance between the Sun and Earth; consequently they are adjusted to reflect the energy received at a distance of one astronomical unit. Unfortunately, daily changes in the Sun's radio flux have only been derived since February 1947, a much shorter period of time than the sunspot index, and so this record cannot be used for long-term studies of solar variability.

As a result, each cycle is usually thought of in terms of the mean relative sunspot number which is computed for each month of its development. Figure 3.2 is a graphical representation of the sunspot cycle between 1749 and 1985 (McKinnon, 1988) as it is defined by the *smoothed monthly sunspot number*. Since the ultimate strength of each spot cycle often varies considerably from others, it is convenient to group all of the previously recorded cycles into categories that

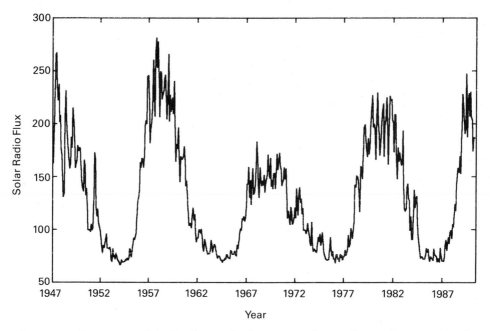

Figure 3.1 The Sun's activity level can also be measured according to its output in the radio frequencies such as at 2800 MHz (10.7 centimeters). This index generally parallels the relative sunspot number except near spot maximum. The diagram depicts this activity between February 1947 and April 1990, after adjustment to a distance of one astronomical unit.

are based upon their intensity at maximum, and to use these guidelines to classify the level of activity for a current cycle.

In a general way, the historical sunspot record can be divided into three such classes: weak (peak intensity less than ninety); moderate (magnitude between ninety and 140); and high (strong cycles with maximum intensities which can be as great as 200 or more). Figure 3.3 shows the differences between representative cycles from each of these categories. The form of the high-magnitude curve is actually more representative of the lower values in its range. Just five recorded cycles have exceeded a maximum of 150, and only two have attained a level of 160 or more. Since four of these cycles have taken place during the past fifty years, the typical sunspot cycle in recent times is one of high intensity.

An examination of Figure 3.3 reveals that regardless of a spot cycle's peak intensity, the time of rise to maximum grows shorter as the ultimate strength at maximum increases. Since the majority of cycles have about the same duration of around eleven years, the strongest cycles are those which are also the most asymmetrical in shape with sharp rises to maximum and slow declines, while the weakest show the opposite form.

This relationship can easily be stated mathematically. According to Vitinskii (1965), the time of rise T_r in years, varies according to

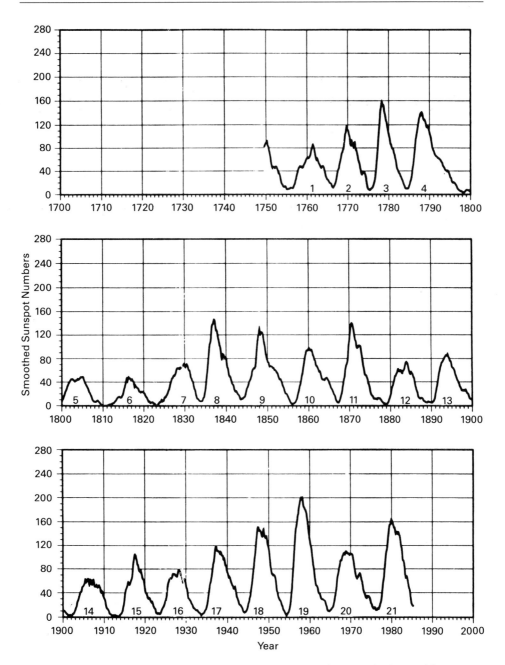

Figure 3.2 Typically, a spot-cycle's level is described by the smoothed monthly sunspot number, a running average which is based upon thirteen conventional monthly-means. The diagram was prepared by Daniel C. Wilkinson of the National Geophysical Data Center.

21

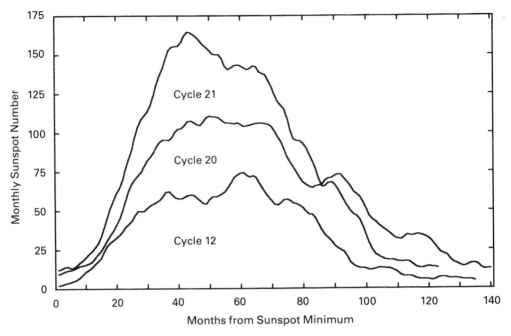

Figure 3.3 Not all spot cycles attain the same maximum intensity. This diagram shows typical examples of low, medium and high strength cycles. Note that the peak intensity occurs progressively later for weaker cycles.

$$T_r \ (0.14 \pm 0.02) = (2.59 \pm 0.10) - \log R_M$$

where R_M is the sunspot number at maximum. Similarly, Waldmeier (1955) finds that the corresponding time to fall to an average minimum value of 7.5, T_f, is given by

$$T_f = (3.0 \pm 0.6) + (0.30 \pm 0.006) \ R_M.$$

This aspect of cycle growth provides one means of predicting the strength and approximate date of maximum for a coming cycle. Since the peak magnitude of each cycle is tied to its rate of growth, the rising branch of a new cycle can be directly compared with a series of mean curves which have been derived from previous sunspot cycles. Waldmeier found that the ascending branches of all cycles, regardless of their intensity, intersect at a common point; near a monthly-mean value of fifty. The succeeding maximum can then be expected to occur some twenty-one months afterward, a time which may be read directly from the appropriate mean curve. This method produces good results, and seems to be limited mainly by a paucity of recorded cycles of very high intensity.

Other popular methods of predicting the future course of a cycle employ a mathematical regression which is also based upon the performance of previously recorded cycles (e.g., McNish and Lincoln, 1949), or relate the coming level of activity to one of several geophysical indices such as the *aa* index (e.g., Schatten *et al.*, 1986). However, at the present time there is no completely reliable way of obtaining long-term predictions of the Sun's activity level.

One can see from Figure 3.2 that even-numbered cycles often have more extended maxima than odd-numbered cycles, and there is some tendency for the heights of consecutive maxima to alternate in strength. The effect becomes even more apparent when the mean intensity of two consecutive odd-numbered maxima is compared with that of the even-numbered cycle which they bracket. Their average magnitude will almost always be higher than the even-numbered cycle. This property is not without exception, but has been strongly in evidence for 200 or more years (Cragg, personal communication).

As this is written (1990), there is considerable discussion as to whether the maximum of cycle twenty-two has taken place. However, the portion of the cycle that has already occurred places it fourth in strength among all recorded cycles, and cycle twenty-one ranks second. Consequently if the odd–even relationship continues, the maximum of cycle twenty-three should also be a very strong one, perhaps with a peak strength which approaches 200. Only one recorded cycle has exceeded this intensity (cycle nineteen which reached a maximum of 201.3 in 1957).

In light of the foregoing information it is apparent that the eleven-year spot cycle is far from regular, at least in the sense that it repeats itself without substantial variation. In fact, many of its parameters encompass a wide range of values. Information from spot cycle number one through a portion of cycle twenty-two (March 1755 to May 1990) is summarized in Table 3.1. A portion of these data were obtained from a similar compilation which appeared in the *Preliminary Report and Forecast of Solar Geophysical Data* (1990).

The number of spots which are visible on the Sun frequently varies greatly over the short-term, and it is common to have low daily sunspot numbers even when the cycle is at its peak. For example, during September 1989, as cycle twenty-two approached its maximum, the daily American Relative Sunspot Number varied between 83 and 284. These fluctuations can be partially accounted for by the tendency of spots to emerge near one another rather than in isolated locations (Liggett and Zirin, 1985), and by the presence of large and long-lived spot groups on the Sun. The latter circumstance results in the appearance of a quasi-periodicity in the number of spots of around twenty-seven days duration (the apparent period for the Sun's rotation), as a large group or active sunspot zone repeatedly rotates onto the visible hemisphere. However, the intensity of a cycle is unrelated to the lifetimes of individual spots, which on the average are a good bit less than a single rotation of the Sun. Moreover, the largest spot groups erupt a year or more before or after the cycle peaks, rather than at its maximum.

Large variations among the monthly averages of Wolf numbers also occur during each cycle. Vitinskii and others have suggested that these values may

Table 3.1

Cycle	Begin	Maximum	End	SSN[a]	Length (yr)	Rise (yr)	Fall (yr)
1	1755 Mar	1761 Jun	1766 May	86.5	11.25	6.25	5.00
2	1766 Jun	1769 Sep	1775 May	115.8	9.00	3.25	5.75
3	1775 Jun	1778 May	1784 Aug	158.5	9.25	2.92	6.33
4	1784 Sep	1788 Feb	1798 Apr	141.2	13.67	3.42	10.25
5	1798 May	1805 Feb	1810 Jul	49.2	12.25	6.75	5.50
6	1810 Aug	1816 Apr	1823 Apr	48.7	12.75	5.67	7.08
7	1823 May	1829 Nov	1833 Oct	71.7	10.50	6.50	4.00
8	1833 Nov	1837 Mar	1843 Jun	146.9	9.67	3.35	6.33
9	1843 Jul	1848 Feb	1855 Nov	131.6	12.42	4.58	7.83
10	1855 Dec	1860 Feb	1867 Feb	97.9	11.25	4.17	7.08
11	1867 Mar	1870 Aug	1878 Nov	140.5	11.75	3.42	8.33
12	1878 Dec	1883 Dec	1890 Feb	74.6	11.25	5.00	6.25
13	1890 Mar	1894 Jan	1901 Dec	87.9	11.83	3.83	8.00
14	1902 Jan	1906 Feb	1913 Jul	64.2	11.58	4.08	7.50
15	1913 Aug	1917 Aug	1923 Jul	105.4	10.00	4.00	6.00
16	1923 Aug	1928 Apr	1933 Aug	78.1	10.08	4.67	5.42
17	1933 Sep	1937 Apr	1944 Jan	119.2	10.42	3.58	6.83
18	1944 Feb	1947 May	1954 Mar	151.8	10.17	3.25	6.92
19	1954 Apr	1958 Mar	1964 Sep	201.3	10.50	3.92	6.58
20	1964 Oct	1968 Nov	1976 May	110.6	11.67	4.08	7.58
21	1976 Jun	1979 Dec	1986 Aug	164.5	10.25	3.50	6.75
22[b]	1986 Sep	1989 Jul	————	158.1	————	2.83	————
Average:				113.8	11.02	4.20	6.73

[a] Smoothed Sunspot Number.
[b] The information for cycle number 22 is based upon a July 1989 first-maximum.

oscillate periodically with an average duration of between three months and one year. Recently, an activity 'period' of about 155 days appeared during portions of the ascending branch of cycle twenty-two, but this varied considerably as the cycle progressed and even disappeared entirely at times. A loose pattern of alternating monthly average intensities has been suspected by others, but its reality has yet to be demonstrated.

A cycle often develops unequally in the Sun's Northern and Southern Hemispheres, and consequently sunspot activity may peak in one hemisphere well before it does in the other. During cycle number eighteen the number of spots in the Sun's Southern Hemisphere reached a maximum in 1947, but the peak was delayed until 1949 in the Northern Hemisphere. Furthermore, a cycle's maximum intensity is frequently much greater in one hemisphere than it is in the other. The Southern Hemisphere dominated during cycle eighteen, but for more recent cycles (including the rise of cycle twenty-two) sunspot activity has been strongest in the north. It is also common for a cycle to alternate between active hemispheres as it develops. These factors seem to vary at random, and do not appear to follow any set rule.

Several investigators (e.g., Eigenson, 1947; Gleissberg, 1958, 1971) have attempted to show the existence of secondary and tertiary periodicities within the historical sunspot record, and even Wolf believed that this was the case. The evidence appears to be strongest for a secondary cycle of about eighty years duration, a cycle which was originally uncovered by Wolf, but has come to be known as the *Gleissberg Cycle* after its main proponent, the famous German astronomer Wolfgang Gleissberg.

Similarly, Schove (1983) and others find some evidence for a 200-year cycle when sunspot data is coupled with that which is inferred from the archival auroral record. The layered sedimentary records known as *varves* (e.g., Williams, 1985) provide some evidence of an even longer cycle of roughly 300 years duration, but the Sun's twenty-two-year magnetic cycle is the most pronounced in this record.

Some of these scenarios adequately reproduce past activity, although none has been particularly successful when their predictive ability is tested. The comparatively short period of reliable sunspot number determinations makes the detection of long-term cycles difficult if not impossible, and the varve records continue to be subject to the interpretation of their cause. Consequently, the eleven and twenty-two year activity cycles are the only period for which the evidence is undeniable.

Substantial irregularities in the sunspot cycle can also occur over longer intervals although it appears that they are rare. The prolonged lull in solar activity called the *Maunder Minimum* which took place between 1645 and 1715 (Maunder, 1922), and the rather poorly defined *Spörer Minimum* (1400–1510?), illustrate this unusual aspect of the sunspot cycle.

Maunder found that between 1645 and 1715 only a few spots were seen on the Sun, and that for sixty years, until 1705, there was never more than one group visible at any one time. Thus it appears that the total number of spots for the entire period would have been less than we have come to expect in any one active year since the lull ended! If experienced observers obtained these observations regularly, it is difficult to understand how the spot cycle remained undiscovered until long after the Minimum ended, since the cycle did continue to operate during this period (Zirin, 1988). However, John Eddy has investigated this aspect of the Sun's activity in great detail and has concluded that it did occur (Eddy, 1976). Eddy managed to unearth a considerable amount of evidence to support the existence of this long period of inactivity, including a dramatic

25

decrease in the number of aurorae and sightings of naked-eye sunspots.

Astronomers are not certain what mechanism could cause such an event. A number of explanations have been suggested, ranging from the interaction of secondary and tertiary sunspot cycles to complex scenarios which deal with the physics of the solar dynamo itself. Eddy summarized the debate over the existence of the Minimum well, when he concluded 'the reality of the Maunder Minimum and its implications of basic solar change may be but one more defeat in our long and losing battle to keep the sun perfect, or if not perfect, constant, and if inconstant, regular ...'.

Although they have appeared briefly at latitudes as high as seventy degrees, sunspots generally occur within thirty-five degrees of the equator. A further interesting property of each cycle concerns the emergence for new sunspot groups within these bands. In 1858, the English amateur astronomer Richard Carrington discovered that the onset of each new cycle coincides with a sudden appearance of new spots at a much higher average latitude than those which erupt near the end of the old cycle. Carrington's conclusions were later con-firmed by Wolf and by Gustav Spörer, who investigated the effect at length. As a result, the process has come to be called 'Spörer's Law.'

At the beginning of the twentieth century, Edward W. Maunder and his wife, Annie S. D. Maunder, thoroughly studied this property and developed the graph which plots the latitudes of emerging spots against time, known as the 'butterfly-diagram.' These graphs are called butterfly-diagrams because of their distinctive shapes, and they provide an important link to Maunder's excellent description of this phenomenon.

Figure 3.4, prepared by Daniel C. Wilkinson of the National Geographical Data Center in Boulder, Colorado, shows the butterfly effect using data which were obtained at the Greenwich Observatory until 1978, and compiled by the National Oceanic and Atmospheric Administration thereafter. The 'leap' in latitude of sunspot emergence which signals the beginning of each cycle can easily be seen around the time of the more recent minimum, which occurred during September 1986.

Since the publication of Maunder's research, astronomers have been aware of an overlap in the sunspot cycle. That is, for several years after the first spots of the new cycle appear, spots from both the old and new cycle frequently exist together. Typically, the first spots from the new cycle emerge about eighteen months *before* the minimum of the current cycle, at a latitude of around ($\pm$) twenty-five degrees. After the old cycle comes to an end, spots from the new cycle begin to erupt at an average latitude which is closer to thirty degrees, and the new spot cycle officially begins. Thereafter the spots tend to appear at increasingly lower locations as the cycle progresses, until the end of the cycle when new sunspots emerge close to the equator, near an average latitude of ($\pm$) seven degrees.

What causes this phenomenon? Although it may originate as rising magnetic

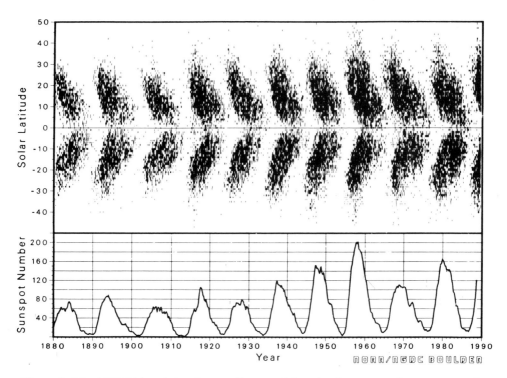

Figure 3.4 In 1858 Richard Carrington discovered that the first groups from a new cycle erupt at high sunspot latitudes and then at gradually decreasing locations as the cycle progresses. This cycle characteristic is clearly shown by E. W. Maunder's Butterfly-diagram. This version of Maunder's plot, prepared by Daniel C. Wilkinson of the National Geophysical Data Center, depicts this activity between 1880 and 1989.

fields are distorted by the Sun's differential rotation (Babcock, 1961), other investigators now suspect that the process is actually a product of 'torsional oscillation,' as described by Howard and LaBonte in 1980. According to this description, spots emerge when a particular latitudinal zone within the sunspot band reaches a maximum in its rotational velocity. If this is indeed the cause, a new cycle would not really begin with the emergence of the butterfly pattern, but would actually be triggered by the onset of a high-latitude torsional wave during the descending branch of the old cycle.

In 1987, Snodgrass and Wilson proposed a mechanism which could prove to be the key of Howard and LaBonte's research. They suggested that huge eddies exist within the Sun's convection zone which enhance the rising magnetic fields by continually forcing them into the Sun's interior so that they become compressed and strengthened. Each convective 'roll' would originate near the Sun's poles, and then gradually move the magnetic concentrations through a network of eddies and towards the equator. Presumably, when the material reaches the normal spot latitudes some eighteen to twenty-two years later, the magnetic fields would be sufficiently intensified so that they would break the solar surface and form sunspots.

If this does prove to be correct, it would indicate that two cycles are in progress at the same time; a result which would be incompatible with the model first described by Babcock. Furthermore, the phenomenon unearthed by Carrington would then be but one manifestation of a more complex process in which sunspots themselves would play only a subordinate role.

In sharp contrast to the Maunder Minimum, recent years have seen the greatest intensities for individual cycles which have ever been recorded and it also appears that the duration of individual cycles has shortened slightly during this time. Are these fundamental changes in the level of the Sun's activity or simply irregularities in the cycle?

Unfortunately, it is too soon to know. If solar activity is increasing over the long-term, then the consequences to the Earth will be considerable, perhaps culminating in a decrease in the size of the polar caps and significant changes in the world's coastline! On the other hand, another long lull in activity would also have serious ramifications. It is possible that climatic conditions could reoccur which would be similar to those that were experienced in Europe during the Maunder Minimum, and the world's food supply might be severly impacted.

Only time can resolve these questions. Even then, it may be that the most predictable of the many aspects of the sunspot cycle is its unpredictability; and perhaps that is where our fascination with the cycle is rooted.

4

The Zurich and Mount Wilson sunspot group classifications

According to astronomer Robert Noyes, 'All sunspots begin their lives as tiny pores no larger than a single solar-granule. They are created when a cluster of sub-surface gas carrying a strong magnetic field breaks the surface ...' (Noyes, 1982). Of course pores are not true sunspots, but as Noyes points out, they are their *precursors*. Many pores are short-lived and last for only an hour or so, while others will go on to become complex sunspot groups which are many times the size of the Earth, and spread over areas which contain thousands of millions of square kilometers.

It is interesting that pores are never found by themselves outside of the Sun's active regions; those areas which contain sunspots, bright faculae or strong fibril structures (Zirin, 1972). Even though there are smaller magnetic structures on the Sun, pores are usually the tiniest features (a few hundred kilometers or so in diameter) which can be resolved with an amateur's telescope.

The Zurich Classification System

In 1938, M. Waldmeier devised the group classifications (Waldmeier, 1947; Kiepenheuer, 1953) which are used by observers who collaborate with the International and American sunspot programs. Although additional techniques are often employed by professional astronomers in special circumstances (the Mount Wilson magnetic classes, for example) this procedure, known as the *Zurich Classification System*, is the method which is used by most of those who observe sunspots visually. Typical examples of these classes are shown in Figure 4.1.

In general, spot-clusters are defined as members of either *unipolar* or *bipolar* groups. A unipolar group is described as a single spot, or a compact cluster of spots, with the greatest distance between two of its members not more than three heliographic degrees. On the other hand, bipolar groups are elongated structures consisting of at least two principal spots which are oriented in a direction that roughly parallels the Sun's equator, but are separated by a minimum of at least three degrees.

Invariably a new cluster first appears as one or more tiny spots without a *penumbra*: the filamented, lighter appearing structure which often surrounds the

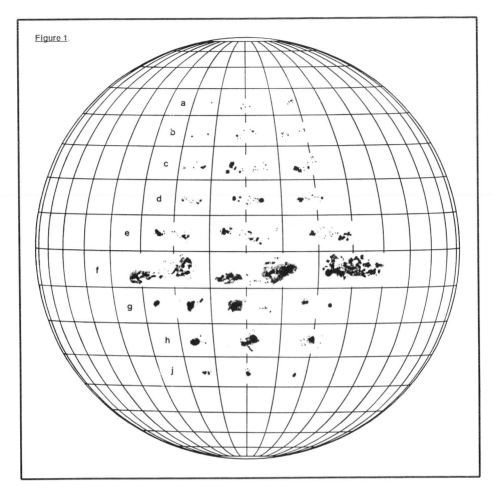

Figure 1.

Figure 4.1 Sunspot groups can be classified according to their evolutionary development. The drawing shows examples from each of the categories of the Zurich Classification System, projected against a modified Stonyhurst disk which indicates their approximate size relative to the Sun's disk.

dark, *umbral* portion of a sunspot. Such a group is classified as type 'A' (Figure 4.1a) according to Waldmeier's system, and does not exhibit bipolarity.

When a small group of spots does show a bipolar structure, it is classified as a 'B' group (Figure 4.1b). Generally, the preceding (most westerly) spot of such a cluster is located at a slightly lower latitude than the following spot. This effect, known as 'Joy's Law,' may play an important role in the exchange of hemispherical magnetic polarity between cycles (Zirin, 1988).

Many groups never develop beyond the A or B stage, and so their lifetimes are usually only a few days at best. Despite their insignificant appearance however, these small activity-centers are often quite numerous, and can make up a substantial portion of the daily sunspot number. Sometimes, and particularly during the minimum of the sunspot cycle, they are the only groups which are seen on the Sun, so the observer must be especially alert for their presence.

A fairly small portion of these immature clusters develop into larger and more complicated spot-systems. The future development of a large group is frequently signaled by a very rapid development at the beginning. Within just a few hours, a tiny new group may grow to become a 'C' or 'D' type cluster. The most distinguishing feature of the C group (Figure 4.1c) is the appearance of a single penumbra around one of its principal spots. The penumbra may surround either the preceding or following spot, but more often than not it first develops around the leader.

The D type group (Figure 4.1d) is characterized by penumbrae which surround both preceding and following spots, although at least one of the spots usually retains a simple structure. Most, if not all, of the class D sunspot groups pass through the C stage as they evolve, but their growth is often so rapid that they appear as A or B types on one day, and as class D on the next. The Zurich system specifies a length of up to ten heliographic degrees[1] for this stage of group development.

A small number of these clusters continue to develop and become E groups (Figure 4.1e). Size and structure are the differences between these two classes. The E group must demonstrate a longitudinal spread of between ten and fifteen degrees, and both preceding and following spots frequently have complicated make-ups. Usually, many small spots are visible between the group's two main spots. The cluster's spread in latitude, which often varies considerably, is not a factor in the classification procedure.

Although E class spot complexes comprise only a small fraction of the total number of groups, a few of them continue to grow larger and become members of the next stage of development in the Zurich system, the type F group (Figure 4.1f). Again, length is the determining aspect. The longitudinal spread for this category of cluster should always exceed fifteen degrees, and it is often much more. These are huge complexes which can be nothing short of spectacular when they are viewed visually.

The number of individual spots in one such group alone often exceeds 100, and they may stretch across as much as one-sixth of the Sun's disk! The lifetimes of F groups are usually quite long, and they frequently rotate back onto the visible hemisphere for second, and occasionally third and even fourth appearances. The greatest flare activity almost always occurs during the time that a cluster is classed as a D, E or F type group.

All sunspot groups have one thing in common: their growth is far more rapid than their decay. A large, complex cluster generally passes through classes A to E in twelve days or less, and then spends the majority of its lifetime in a slow

---

[1] Near the Sun's center, ten degrees is approximately one-twelfth of its apparent diameter.

disintegration process (Bray and Loughhead, 1965). A group's maximum growth can frequently be recognized by a lack of continued spread in longitude. When the many small sunspots which are characteristically associated with developing clusters start to disappear, it is almost always a sure sign that decay has begun.

The decay process is similar for all spot clusters. The smaller C and D class groups usually dissolve into a rather long-lasting single spot with penumbra. The larger D, E or F class complexes frequently decay into two spots, each with penumbrae. A decayed bipolar group of this kind is classified as type G (Figure 4.1g). These clusters have no small spots between their primary spots and exceed ten degrees in length. They generally continue to dissolve into a single spot which is most likely to be the proceding one.

If decay does terminate in a single spot, the cluster is assigned to class J when it is less than two and a half degrees in diameter (Figure 4.1j), or H, if its longitudinal spread is greater than this amount (Figure 4.1h). The H class spot which remains from the decay of an F group is particularly long-lasting, and may return to the visible hemisphere for several successive solar rotations. A few small spots occasionally erupt around the boundaries of these groups, although according to the National Oceanic and Atmospheric Administration, when strong new spots emerge near a mature unipolar cluster, they are likely to be members of a separate group (*Solar-Geophysical Data*, 1986).

Professor Waldmeier (1955) has provided the following rule-of-thumb for determining the approximate lifetime of a spot cluster in days, t, according to its area:

$$t = 0.1 \, A_{\text{Max}}.$$

In this case, A_{Max} represents the maximum area attained during the growth process, expressed in millionths of the Sun's visible hemisphere; an amount which can be estimated according to a method which will be described in a later chapter. The result which is obtained through the application of this relation should be regarded as approximate.

All of the spot-clusters that we have described (with the exception of types A, H and J) have visually bipolar structures. Occasionally a large group develops which lacks this composition. Clusters of this type are included in class H even though they do not represent a decayed group. Usually they grow to be quite large and may become prolific flare producers. These groups frequently carry both magnetic polarities within a single penumbra, and so they are said to be *magnetically bipolar*. Less often, a small unimpressive group without a bipolar structure will occur, and it should be classified J, even though it too did not come about through a dissolution process.

When Rudolf Wolf devised the concept and method of computation of the relative sunspot number, he envisioned an evolutionary system of sunspot growth. The usefulness of the Zurich classification is also rooted in its developmental nature. Consequently, when a spot cluster is assigned to a specific category, an investigator has a reasonably accurate indication of its future development and eventual fate. The information which is gained in this way has importance for studies of the growth and decay processes of spot groups, and when planning specific observing programs.

In addition to the basic sunspot group categories that we have discussed above, and which constitute the Zurich system, more detailed descriptions have recently been devised which are provided in the following section. These are especially important to those who are interested in the relationship between a group's characteristics and its production of solar flares and other of the Sun's active phenomena.

Sub-classifications of sunspot groups

In 1973, the well-known solar physicist and astronomer, P. S. McIntosh, extended the existing Zurich classification scheme so that it would more accurately reflect the conditions within each group. In the McIntosh revision, several of a group's structural characteristics are represented by a series of three uppercase letters. The first of these components embodies the spot-cluster's type according to the 'modified' Zurich classification.

When spot-clusters are categorized according to the modified system, G class groups are included within the definition of classes E and F, and those which are members of the J category are incorporated in class H. The modified system is regularly employed by the Space Environment Services Center, the sister organization of the National Oceanic and Atmospheric Administration, and by others who are engaged in specialized studies of the Sun's phenomena.

Dr McIntosh devised the following sub-classifications for sunspot groups which are expressed by the second and third letters of a three-letter code (*Solar-Geophysical Data*, 1986). A complete description of the rationale behind the McIntosh system is provided in McIntosh (1981).

Second-letter codes

The second letter indicates the type of penumbra (if any) which surrounds the largest spot of a group.

> X: No visible penumbra which exceeds three arc-seconds in width.
> R: A simple, or rudimentary penumbra. That is, the penumbra is incomplete or irregular.
> S: A symmetrical penumbra. The penumbra is circular in appearance, and does not exceed two and a half degrees in diameter.
> A: An asymmetrical penumbra. The penumbra is oval or complex in appearance, with a diameter which does not exceed two and a half degrees of longitude.

33

H: A large circular penumbra with a diameter larger than two and a half degrees.

K: A large oval penumbra with a diameter which is larger than two and a half degrees. When these penumbrae exceed five degrees in longitudinal spread they almost always contain *both* polarities within the single penumbra, and are categorized as members of the Mount Wilson 'gamma' magnetic class (see the following section). Sunspot groups of this kind are frequently active flare producers, and along with the delta groups are the most likely types to be associated with solar 'white-light' flares (Neidig and Cliver, 1983).

Third-letter codes

The third letter of the code describes the distribution of individual spots within the sunspot group.

X: A single, isolated spot.

O: An open spot distribution. The area between the principal preceding and following spots is free of individual spots.

I: An intermediate distribution of spots. In this category, a few individual spots lie between the principal preceding and following spots.

C: A compact spot distribution. Numerous small spots exist between the main preceding and following spots, and at least one of them possesses a penumbra. This distribution is also associated with high rates of flare production.

For example, under the guidelines of these sub-classifications, a sunspot group with the three letter definition EKI, would describe a spot group with the following characteristics:

A visually bipolar cluster with an east to west spread of between ten and fifteen degrees. The group has one large oval penumbra which fully encloses at least one of its principal spots, and a second penumbra surrounding its other main spot. Several small distinct spots are visible between its preceding and following spots.

The Mount Wilson system

The instrumentation which is required to carry out measurements of the Sun's magnetic structure in accordance with the *Mount Wilson classification system* is generally not available to amateur astronomers. Consequently, the following short explanation of its magnetic classes (*Publications of the Astronomical Society of the Pacific*, 1947; Abetti, 1961; Cragg, personal communication) is included for the reader's convenience, rather than for observational purposes.

The Mount Wilson and Zurich systems are not rivals: each plays a specific role in our knowledge of sunspots and the solar cycle. For their part, the Zurich classes continue to play an important role in statistical studies which deal with solar flares, and in research of the evolutionary growth and decay of spot groups. However, much of our understanding of the Sun's dynamic phenomena and interior structure has come from studies of the magnetic character of sunspots. It is important that the reader not be misled by the visual appearance of a spot cluster in this regard. The structure of a group, as it appears when viewed visually, is not (necessarily) indicative of the cluster's magnetic qualities.

Since all sunspots are descended from the relatively tiny magnetic features on the Sun known as pores, they too are associated with strong magnetic fields. The dominant magnetic polarity for the principal preceding spot of bipolar groups in the Northern and Southern Solar Hemispheres is almost always opposite. This quality begins to reverse for the new sunspot cycle at about the time that the old cycle reaches its maximum intensity, in accordance with the Hale-Nicholson Law of hemispherical polarity (Hale *et al.*, 1919). Consequently, the polarity's magnetic sign returns to its original value at twenty-two year intervals, a period of time which is referred to as the Sun's magnetic cycle.

According to Leighton (1964), the process that guides this change of hemispherical polarity occurs as a result of *Joy's Law*, an effect which describes the angular relationship between spot complexes and the solar equator. Joy found that the preceding (most westerly) spot is almost always located nearer to the equator than the following spot, a characteristic that becomes more pronounced at higher latitudes. The angle varies from about three degrees for groups which are positioned close to the equator, to around eleven degrees for those at a latitude of ($\pm$) thirty degrees (Hale and Nicholson, 1938).

When groups decay, fragments of their magnetic concentrations are carried towards the Sun's poles by a flow of gases within the granular network. Since the polarity from the trailing spot reaches the polar region sooner (remember that it is slightly closer at the beginning) and is of opposite sign to that of the old preceding spot, the process eventually leads to a switch of polarity between the Sun's Northern and Southern Hemispheres. Generally this reversal, or 'magnetic-rollover,' occurs around the time of sunspot maximum. During sunspot cycle number twenty-two, the polarity of the preceding spot for Northern Hemisphere spot groups will normally be negative, and that for the following spot will be positive.

The Mount Wilson scheme for categorizing sunspot groups is one of the most scientifically defined methods of sunspot classification. The system was developed in the early part of the twentieth century by G. E. Hale and others at the famous Mount Wilson Solar Observatory in Southern California (Hale *et al.*, 1919). The dynamic Dr Hale (Hale's degree was honorary; he was never able to find the time to earn a doctorate) is considered by many to be the 'father' of modern solar astronomy (Howard, 1968).

Hale devised his method so that it would incorporate the position of a spot cluster within its associated plage, or faculae, after he realized that localized magnetic fields existed within these areas. Initially, the measurement techniques at Mount Wilson were somewhat deficient in that they were of low resolution and showed only the dominant polarity of each spot group. However, this early difficulty has been overcome with modern instrumentation, particularly by the invention of the magnetograph by Harold Babcock in the 1950s (Babcock, 1953). Consequently, in spite of being defined over seventy years ago, the Mount Wilson system continues to be successfully employed today.

Hale's classification scheme employs the first three letters of the Greek alphabet, α (alpha), β (beta) and γ (gamma), to classify groups according to a spot's magnetic, rather than visual characteristics. The differences between these categories are depicted in Figures 4.2 and 4.3, and are outlined below.

An alpha-type group is defined as a single unipolar spot, or small cluster of spots, with the *same* magnetic polarity. In practice, however, a few bits of opposite polarity are occasionally found to be present within these groups. Bray and Loughhead (1965) suggest that these few exceptions to the unipolar character of a cluster are insignificant in strength and effect, and consequently they can be ignored when classifying the group.

This (unipolar) category is divided into the following separate classes according to a spot's location relative to its associated faculae:

α: The faculae which leads and trails the spot is symmetrically distributed.

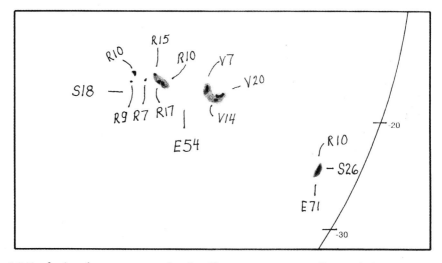

Figure 4.2 Professional astronomers also classify spot groups according to their magnetic characteristics. Typical examples of the Mount Wilson system's alpha and beta-type groups are shown in this figure. Note that in the beta-cluster (left), all leading or trailing magnetic concentrations are of the same sign while only a single polarity is represented in the alpha group.

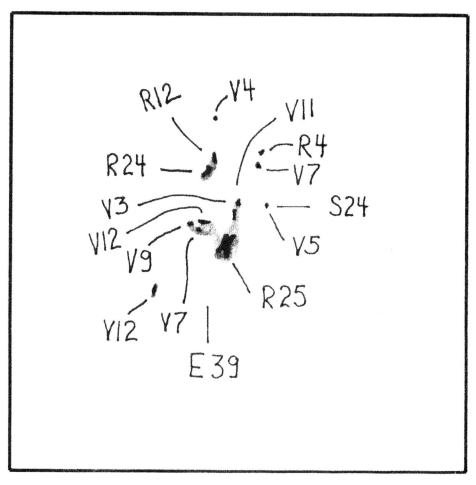

Figure 4.3 According to the Mount Wilson system, when the polarities of the magnetic concentrations within a group are randomly mixed, the spot cluster is given the gamma classification. Figures 4.2 and 4.3 are reproduced from drawings made at the Mount Wilson Observatory.

αp: The faculae which follows the spot group is of greater area than that which precedes it.

αf: The faculae which precedes the cluster is of greater area than that which follows it.

On the other hand, beta groups are those in which distinct preceding and following sunspots, with polarities in magnetic balance, are present. That is, a cluster where at least two principal spots (or two main magnetic concentrations) of *opposite but equal* polarity are clearly associated. In a typical bipolar group, this usually takes the form of a string of individual polarities which extend across the group, but separate into opposite polarities near their center. The simple, magnetically bipolar spot cluster forms the core of the Mount Wilson system: Hale and his cohorts at Mount Wilson felt that all other clusters were variations of this fundamental group-type.

Beta clusters are divided into the following sub-categories:

β: The preceding and following spots are of equal strength.

βp: The preceding spot is the dominant member of the group.

βf: The following spot is the dominant member of the group.

βγ: The magnetic qualities of the group are not clearly divided, or are slightly mixed.

In the majority of spot groups (Hale found around fifty-seven percent), the preceding spot is the dominant and longer-lasting member. Zirin (1988) attributes this effect to the relative rarity of clusters containing several large single dipoles, and finds that even when a group's polarity is irregular (see below) the leading spot is still usually larger than the trailer.

A gamma-type group is difficult to distinguish from an alpha cluster without a special knowledge of its magnetic polarities, and is not well represented among the Zurich classifications. However, it is defined as a group without beta-type concentrations, but with *mixed* polarities within a single penumbra. These clusters frequently have significantly larger areas than those of the alpha group. The gamma groups are not generally called γp or γf, but instead are referred to simply as gamma-types. Figure 4.3 illustrates a typical example of this class in which the polarities are completely mixed.

If the magnetic polarities are normal, or 'regular' for solar hemisphere and cycle, the examples depicted in Figures 4.2 and 4.3 are correct. However, it is entirely possible to have (for example) an αp group which carries the polarity of the follower, just as it is possible to have an αf with preceding polarity. In these situations, the classification is determined by the group's position within its plage, and a note is added which states that the polarity is irregular. In this way an αf with preceding polarity would be simply described as, 'αf with irregular polarity.'

Although these definitions are correct in the strictest sense, there is some tendency to classify p or f from the field-strength alone, especially when analyzing the beta-type complexes. Most of the time the two approaches agree because there is a good correlation between field-strength and sunspot area, but this is not always true (Cragg, personal communication).

In any cycle, roughly two and a half to three percent of the groups show a reversed, or irregular polarity. They appear to be normal spot clusters in every other way, but differ in that the preceding portion of the group has follower polarity and the trailer has preceding polarity. Occasionally a complex group with reversed polarity will demonstrate a high level of activity. However, statistical studies (e.g., Richardson, 1948) show that these groups do not have any other outstanding features which set them apart from their brethren; they appear to be just as strong, and only slightly less-enduring than the similar stage of a group with normal polarity.

A final category, the δ (delta) configuration, was added to the Mount Wilson system by researchers at the National Oceanic and Atmospheric Administration. It is appended to any basic group class when spots of opposite magnetic polarity are located within two degrees of one another within the same penumbra. The delta classification has been included in the Mount Wilson system because of the very high incidence of solar flares which is associated with this particular configuration. In this way, attention is called to groups with a high probability for flare production, some of which are particularly likely to result in strong terrestrial effects. The more noteworthy of these consequences are discussed in the following chapter.

5

Common terrestrial effects

Nearly fifty years before Schwabe discovered the sunspot cycle, Sir William Herschel wrote: 'The influence of this eminent body on the globe we inhabit is so great and so widely diffused that it becomes almost a duty to study the operations which are carried on upon the solar surface ...' (Todd, 1899). In spite of Herschel's later (and fruitless) attempt to tie the price of wheat to the Sun's spottedness (Menzel, 1949), his early insight into the connection between events which occur on the Sun and those in the terrestrial environment was quite remarkable. This is especially true when one considers that the solar wind and flares were unknown at the time, and the real nature of the Sun would not be uncovered until much later.

While we are all aware that life (at least as we know it) could not be possible without the Sun's effects, it also influences the Earth in ways which while beautiful to behold, can also be disruptive and dangerous. Fortunately, almost all of the Sun's potentially harmful electromagnetic radiation is blocked or dissipated in the Earth's magnetic field and atmosphere.

The geomagnetic field is actually a combination of magnetic fields which originate in three principal areas: the main field which is internal to the planet and varies slowly; fields which change according to external influences such as variations in the upper atmosphere; and those that result from currents flowing within the Earth's crust.

The main internal field gives rise to a comet-shaped region around the Earth which normally extends outward towards the Sun for a distance of around ten Earth radii, and trails the planet tail-like for a thousand radii or more. This feature, known as the *magnetosphere*, forms a natural barrier between the Sun and much of the Earth's surface, although it does not protect the daylight side of the polar regions as well as other locations. But perhaps we are getting ahead of the story, which really begins with the 'solar wind.'

The Sun continuously emits a high-temperature (approximately one million kelvin) stream of protons and electrons from the corona and these travel outward at a speed in excess of 300 kilometers per second. This electrically conductive plasma is known as the *solar wind*, and it plays an important role in the solar–terrestrial relationship by combining with the Earth's field to create the magnetosphere.

The electrically charged plasma particles cannot move through the geomagnetic field to the atmosphere; thus, the solar wind blows mainly around the Earth, and this process forms the magnetosphere. However when this happens, the wind moves rapidly by magnetic fields which result from an interaction between those carried by the wind and the Earth's field (Akasofu, 1982). A few of these fields interconnect across the boundary between the magnetosphere and the wind where their pressures are in balance (the *magnetopause*), and as it blows across the interconnection, the boundary becomes a powerful electric generator. The currents which are generated by this process are subsequently discharged through the atmosphere.

The geomagnetic field lines meet the Earth in large circular areas which are centered on the magnetic poles. Through a complex process, the power of the generator is increased by the incoming electrons and they are accelerated along the converging field lines towards the upper atmosphere at the poles. Eventually they collide with, and ionize atoms and molecules in the ionosphere and produce aurorae.

Since the generator is powered by an ever-changing solar wind speed and field intensity, its overall strength also varies. Moreover, strong solar flares can cause gusts in the wind which create giant shock waves as they chase down and contact the normal, slower wind flow. When these shock-fronts reach the magnetosphere, the power of the generator is greatly increased, and the electrical discharge which follows frequently produces an intense magnetic disturbance.

However, flares are not the only solar phenomenon which cause intense magnetic storms. Long-lasting, low-density areas of the Sun's upper atmosphere called *coronal holes*, also produce 'gusts' in the wind which have been shown to cause recurrent geophysical storms (e.g., Sheeley and Harvey, 1981). The fronts of the gusty particle streams from these phenomena contain strong magnetic fields which can also increase the power of the generator. The resulting magnetic disturbances reoccur on a time-scale of about twenty-seven days (the period of apparent solar rotation) as a hole repeatedly returns to its original position relative to the Earth; an effect which lasts for several days.

Furthermore, it is now believed (Hewish *et al.*, 1985; Hewish, 1988) that coronal holes may be responsible for *all* geomagnetic disturbances, and that flares and other active solar phenomena only intensify these effects. At the very least, the holes play a critical role in the acceleration of the solar wind and some interaction between flares and coronal holes appears to be necessary (Thompson, 1989).

The magnetosphere absorbs and deflects away mainly those particles which carry an electrical charge, and so the Sun's X-ray and ultraviolet radiation can pass through it relatively unhindered (Noyes, 1982). But these rays soon encounter the outer layers of Earth's atmosphere where most of their energy is expended through interactions with its basic components. Eventually the

41

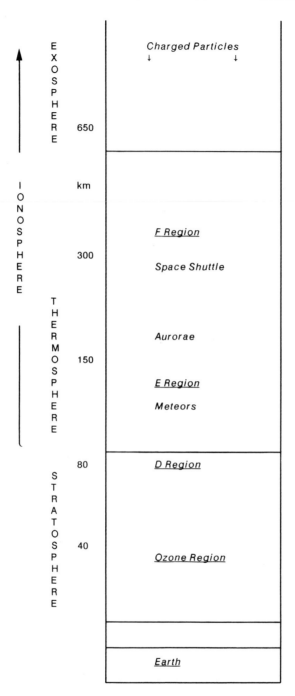

Figure 5.1 The principal layers of the Earth's atmosphere. Since spacecraft in Earth orbit such as the Space Shuttle circle the planet at relatively low altitudes, the atmosphere offers them a reasonable amount of protection from many of the more harmful of the Sun's rays.

remainder descends into the ozone and stratospheric layers which absorb all but a little of the remaining radiation, leaving just enough to support life on the Earth (Figure 5.1).

In spite of these impediments, an especially strong surge of activity can have far-reaching consequences, wherein even the magnetopause is compressed up to half of its normal distance from the Earth. As a result of effects such as these on the Earth's environment, radio and other communications media are subject to disruptions and blackouts, satellites and other spacecraft can be affected, and compass errors near the poles and elsewhere make local aircraft navigation difficult or impossible.

The effects from events which originate on the central to western portion of the Sun are felt at the Earth somewhat sooner than those from more easterly locations since these particles are able to travel to the Earth on a more direct path than their counterparts (Figure 5.2). The characteristic shape of this field, known as the *Archimedean Spiral*, is caused by the Sun's rotation. (The Sun rotates nearly sixty degrees westward during the four and a half days which it normally takes for the solar wind to travel to Earth.)

Geophysicists believe that the magnetic field is substantially the same today as it was many millions of years ago, so some of these influences were undoubtedly experienced by the earliest creatures to traverse our planet. Since our knowledge of the processes which cause these conditions has improved considerably during the last few years, it has become possible to forecast the occurrence of some of them. This is accomplished by monitoring changes in the solar wind with instruments aboard the International Sun-Earth Explorer satellite (ISEE), although due to the speed of the Sun's emissions, only an hour or so of warning time is currently available.

The indices of geomagnetic activity are among the important tools which atmospheric and other scientists use to measure variations in the magnetic field which result from irregular electrical flows. These records gauge the varying electric currents in the terrestrial environment with instruments which are called magnetometers; devices which measure variations in the intensity and oscillation of magnetic forces.

For the most part, the equipment operates at a global array of mainly Northern Hemisphere magnetic observatories located at sites which are chosen for their latitude and equal distribution. Other instruments operate from spacecraft high above the Earth such as the Geostationary Operational Environmental Satellites (GOES), while changes in the interplanetary magnetic field are monitored by a magnetometer aboard the Pioneer Venus Orbiter.

The *K-index* is one of the more valuable of the geomagnetic indices. It is derived according to a scale which varies from zero (very quiet) to nine (extremely disturbed). The K-index is measured by each observatory at three-hour intervals, and is then standardized so that it has about the same base-line at all sites. Its main use is to provide a foundation for the whole-Earth, or *planetary*

43

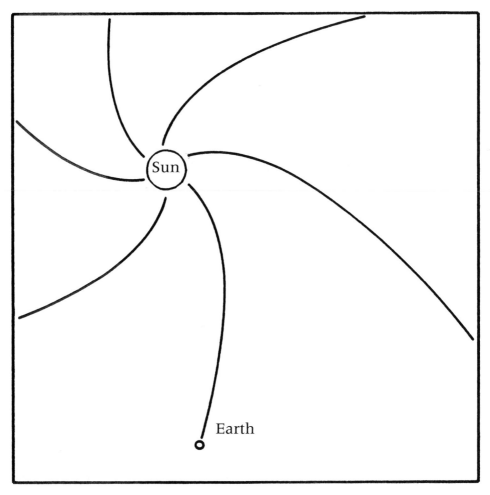

Figure 5.2 Since the Sun rotates on its axis, the paths of solar magnetic fields carried outward by the solar wind *are twisted into a pattern which is known as the Archimedean Spiral.*

K-index, 'K*p*;' an average of the standardized K-indices which have been determined by all contributing stations.

However, since large-intensity geomagnetic disturbances frequently attain their maximum level over intervals which exceed one day, a separate index known as the 'A*p**' index was devised in 1978 (Allen, personal communication). According to this system, a major storm is defined by an A*p** level which equals or exceeds forty, and more than a thousand of these storms occurred between 1932 and 1989. The relationship between the sunspot cycle and the number of events which have been recorded during each year of this interval is shown in Figure 5.3. It is interesting to note that the number of these disturb-

Table 5.1

Activity Level	K-Index
Quiet	Usually no indices above 2
Unsettled	Usually no indices above 3
Active	A few indices of 4
Minor storm	Indices mostly 4 or 5
Major storm	Some indices 6 or greater
Severe storm	Some indices 7 or greater

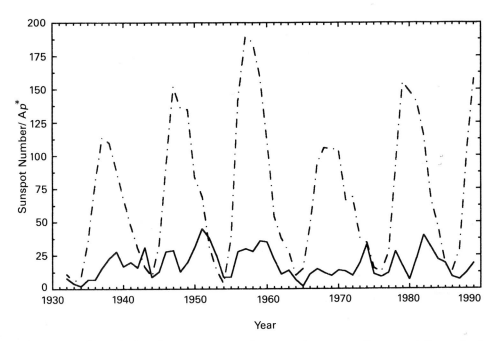

Figure 5.3 The majority of great geomagnetic storms do not usually occur at the peak of the spot cycle, but rather take place during its descending phase. This activity, as it is described by the Ap^* index between 1932 and 1989 (solid line), is shown plotted against the smoothed monthly sunspot number.

ances reaches a peak during the declining phase of the cycle when the number of coronal holes is also greatest (Akasofu, 1982).

The degree of magnetic storm activity can also be associated with the value of the K-index. One such system, employed by the Space Environment Services Center in Boulder, Colorado, ranks the level of magnetic disturbance according to the scale shown in Table 5.1 (*Preliminary Report and Forecast of Solar Geophysical Data*, 1987).

When the Kp-index exceeds a value of six or so, there is a chance that one of the best known of all the extraordinary effects of the Sun's activity upon the Earth, the *aurorae*, can be seen at middle and lower latitudes as well as in the higher auroral zones (Bartels and Chapman, 1957). Aurorae, known commonly as the 'northern (or southern) lights,' have been recorded since Aristotle first wrote of them in the fourth century BC.

Although some have attributed the term 'aurora' to Galileo, it is likely that Northern Hemispheric aurorae were first referred to as the *aurora borealis* (northern dawn) by the French scientist Pierre Gassendi in 1621 (Petrie, 1963). On the other hand, aurorae which occur in the Southern Hemisphere were named *aurora australis* over a century later, by Captain James Cook who observed them at sea during his historic voyage in the latter part of the eighteenth century.

Since these events can be very spectacular, and at the same time appear suddenly and without warning to the uninitiated, a number of legends have arisen in connection with them. For example, the Vikings believed that the 'lights' were flashes from the shields of warriors bound for a heavenly Valhalla; Eskimos imagined lights held in the hands of dangerous spirits which led the dead to new lands of plenty, and the more pragmatic Swedes thought the effect was caused by the torchlight of Lapps stalking stray reindeer. These legends are not confined to people who inhabit high-latitude areas: Petrie describes an early aboriginal belief which viewed the lights as gods dancing across the night-time sky, and the Ceylonese connected their appearance with Buddha's displeasure.

In spite of having been recorded for more than 2000 years, it has only recently been discovered that aurorae are actually a discharge phenomenon in the upper atmosphere which occurs as electrical currents enter the atmosphere from the magnetopause, flow along the polar ionosphere, and return. When the charged particles collide with atoms and molecules in the upper atmosphere these components are excited (raised to a higher energy state) and begin to radiate, or glow, as they return to normal, resulting in an aurora. The procedure is similar to that of a neon light.

Since the particle stream is directed towards the Earth's polar regions, aurorae are much more common at high latitudes. They occur most frequently in two huge 'auroral ovals,' which are centered near the magnetic poles. In the Northern Hemisphere the zone passes over Alaska, Greenland, Scandinavia and Siberia (Akasofu, 1982). On occasion though, an especially intense burst of energy such as that produced in a strong flare, causes the power of the magnetospheric generator to increase dramatically; then the oval enlarges and aurorae can also be seen at lower latitudes.

Broadly speaking, the form of aurorae are either rayed, non-rayed or irregular. The characteristic shape of the auroral 'arch' is initiated by the field which controls the path of the inward spiraling electrons, although its visual appearance is somewhat dependent upon the height at which a particular effect occurs, and by variations in the solar wind and magnetosphere. The arches are eventually broken into rayed and curtained effects by the continued impact of the stream on the upper atmospheric constituents (Noyes, 1982).

Both Northern and Southern Hemisphere aurorae generally take place between 100 and 400 kilometers above the Earth's surface, although some rare forms occur at heights of 1000 kilometers or more. The prominent green homogeneous arcs and fast-moving rays usually originate at an altitude of between 110 and 250 kilometers. Others, the fainter arcs and reddish patches for example, are most active at a height which is closer to 300 or 400 kilometers. Their brightness typically ranges from ten to 100 times the normal night-sky level.

Aurorae display a wide variety of shapes and colors, but they are most often a pale, greenish-white. Their color is primarily a function of the excited states of the atoms and molecules which are impacted by the electron stream. This is seen mainly in the emission of green oxygen spectra at 5577 ångströms[1] (Å) and in red spectra at 6300 Å and 6363 Å. A less-frequent emission in the blue and violet region usually results from interactions with atoms and molecules of nitrogen (Noyes, 1982).

The geomagnetic storms which are associated with aurorae are most apt to occur around the Spring and Fall (autumn) equinoxes when the Earth is in a position which is more likely to intercept the Sun's highly directional emissions (Figure 5.4). Since the growth of large sunspot groups beneath coronal holes seems to suppress aurorae, they appear more frequently in high latitudes during the declining portion of a spot cycle. On the other hand, flare-generated aurorae are more common near the maximum phase of the sunspot cycle when the number of solar flares is also greatest. Even so, the most spectacular of the low-latitude events often occur a year of so before or after spot maximum when the greatest of the solar flares erupt (Zirin, 1988).

When aurorae are seen in unusually low latitudes their description is especially important to the understanding of the terrestrial environment. An average of less than five of these intense magnetic storms can be expected each year (Giovanelli, 1984). A number of amateur organizations compile information about aurorae. The Division relays the information to National Oceanic and Atmospheric Administration for use by researchers working within a number of disciplines.

Unfortunately not all of the effects of the Sun's emissions influence the Earth in the benign way of an aurora. When electrical currents of sufficient strength flow in the atmosphere, their charges can be transferred to ground level and into long power-lines and telephone lines. Many of the older power systems are especially vulnerable in this respect, and may experience intense fluctuations and component burnouts. During the great magnetic storms which occurred during October 1989, millions of Canadians were without power for up to nine hours and damage to power distribution systems ran into millions of dollars. The

---

[1] An ångström is the unit of measurement for a wavelength of light. The width of one ångström is one hundredth of a millionth of a centimeter.

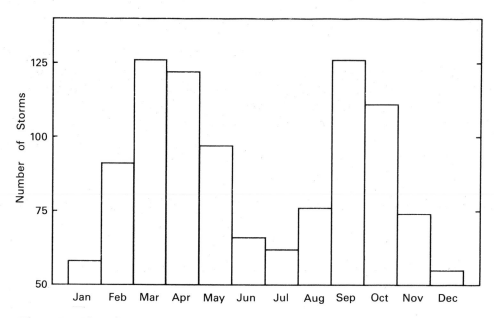

Figure 5.4 Since the Sun's emissions are highly directional, the resulting geomagnetic disturbances occur more frequently near the Spring and Fall equinoxes when the Earth is in the best position to receive them.

currents can also flow along pipelines and cause electrolytic reactions which corrode the metal pipe.

The disruption of radio and other communications facilities is another important consequence of intense geomagnetic activity. The most severe of these difficulties begin soon after the initial outburst of an especially intense flare. With the exception of the ISEE satellite, special X-ray telescopes and other instruments aboard satellites such as the GOES spacecraft are usually the first to detect these conditions. Virtually immediately thereafter, the radiation begins to affect the ionosphere.

The *ionosphere* is a low-pressure area of the atmosphere which is composed of charged atoms, molecules and electrons, and begins at a height of about seventy kilometers above the Earth. For our purposes, it can be thought of as being composed of several separate areas at increasing heights, the D, E and F layers. The F-layer is the principal means of propagation for short-wave radio signals, while the E-region is associated with the transmission of the normal broadcast frequencies. The lower D-layer absorbs some waves and reflects others, mainly those of low and very low frequency. (When the D-layer is sufficiently enhanced by the ionizing effects of charged-particle interaction, it is possible to monitor the frequencies which are propagated by it, and detect many of the Sun's flares indirectly. Chapter 11 contains information on this technique.)

After the effects of a strong flare reach the D-region, powerful electrical currents begin to flow in the ionosphere, and its capability to reflect radio signals is altered. Some of the signals are intensified and propagated over unusually long distances, but others are absorbed and lose their energy, resulting in a weakening or loss of the affected medium. Occasionally, the effect is so intense that high-frequency radio communication becomes impossible. At the poles, where the consequences are significantly greater than at other locations, the disruption can last for days.

Satellites and other spacecraft are also influenced by the radiation from flares. Their solar panels can deteriorate very rapidly, the satellite's course and position can be altered, or they may go awry in other unforseen ways (Heckman, 1988). This is an especially serious problem with navigational satellites, since ships at sea rely upon these instruments being in precise positions in order to determine their locations with a high degree of accuracy. During the severe disturbances which occurred during the Fall of 1989, about ten percent of the objects in space were temporarily lost due to flare-related effects.

In addition to these effects, the flowing electrical charges can build up in a satellite's electronic equipment and cause it to malfunction. Or, if the spacecraft encounters a high-density plasma cloud from a strong flare, the resulting discharge of thousands of volts can quickly destroy its delicate electronics circuits. Just one such particle can penetrate the satellite's computer memory-circuit and cause the computer to reset or to issue false commands. The latter type of disruption is frequently referred to as a *single-event upset*.

Normally, X-radiation is not a severe problem for low-orbit spacecraft such as the Space Shuttle, since these vehicles operate at an altitude which is near enough to the Earth so that they are reasonably well protected by its atmosphere. However, the radiation produced by flares will almost certainly pose a threat to astronauts who venture beyond the natural protection which the atmosphere provides, and special shields will have to be applied to the spacecraft which will carry men and women to the planet Mars. (It has been estimated that an especially energetic flare could subject astronauts outside of the Earth's environment to a lethal amount of radiation within several days.)

Powerful emissions from the Sun can also cause dramatic changes in a different arena. The large increase in ultraviolet radiation that often accompanies a strong flare can cause the Earth's *thermosphere* (the atmospheric region which is most affected) to increase in temperature and expand. When this occurs, the atmospheric density at a given altitude increases, and a satellite orbiting near that height encounters more atmosphere than it would otherwise. The increased resistance slows the satellite and it descends at a faster-than-anticipated rate.

During March 1989 a similar set of circumstances occurred (M. Taylor, personal communication), resulting in a thermospheric expansion which caused the old Solar Maximum Mission (SMM) spacecraft to 'drop' in altitude by nearly a full kilometer! Satellites in geosynchronous orbits, some 36 000 kilometers above Earth's surface, are usually not affected by this process, however.

Some expansion and contraction of the Earth's atmosphere is a normal consequence of the annual changes in the Sun-to-Earth distance, and may also result from small changes in the Sun's luminosity. The Sun's total energy output, the solar 'constant,' is measured by radiometers and other instruments aboard spacecraft such as the SMM, and at observatories on Earth. As one might expect, the Sun's output actually decreases during the passage of a large sunspot group across the visible disk (Noyes, 1982).

Recently, however, it has been suggested (Willson and Hudson, 1988) that over the long-term, the amount of energy which emanates from the Sun may instead parallel the waxing and waning of each spot cycle. If this does prove to be correct, and is combined with the large increase in flare activity which is normally associated with cycle maximum, the effect upon a satellite's orbit could be devastating. In fact this scenario appears to have already occurred: the expansion of the atmosphere caused by the effects of an unexpectedly strong surge of solar activity was the principal cause of the early, and disastrous demise of the Skylab spacecraft in 1979.

The intense activity which has accompanied the rise of sunspot cycle twenty-two has resulted in a repetition of this situation. What was once only a threat to the SMM spacecraft has become reality with the fiery demise of this extremely productive instrument during November 1989. The only solution to the degradation of a satellite's orbit which results from these processes is to move it higher to a thinner (and consequently less-protected) part of the atmosphere, or to bring the more important instruments back to Earth with the Shuttle.

Do small changes in the Sun's output influence our weather and climate? Most of the statistical connections between the sunspot cycle and the Earth's weather have evaporated over time. Those relationships which do appear, are often complex and seem to vary considerably according to geographical location and time (Noyes, 1982). Frequently what appears to be a valid correlation at one site will show a mixed or even opposite relationship in a nearby area.

For example, an examination of the climatic record between 1880 and 1960 has disclosed a minimum of rainfall in Cairns, Australia, and a maximum at Hobart, Tasmania. Both appear to be well-correlated with the sunspot cycle for reasonable periods of time, but with associations which are almost exactly opposite of one another! The apparent lack of a physical mechanism which could account for changes of this magnitude is even more troubling. Moreover, the effects themselves are generally thought to be too weak and short-lived when compared with traditional weather patterns to produce changes on the order of those which are observed. An extremely interesting discussion and non-technical analysis of these relationships is provided in Giovanelli (1984).

One outstanding instance of a relationship between solar activity and terrestrial weather that has withstood the test of time is the drop in temperature (approximately one degree centigrade on the average) which coincided with the Maunder Minimum (Maunder, 1922). During this prolonged 'lull' in the Sun's activity, many parts of Europe suffered through a period of extremely cold and dry weather, in what has come to be called a 'little' ice age. Although a change of only one degree centigrade appears to be almost insignificant, the temperature decrease was maintained over a long interval, and the combination of time and lowered temperature is the likely cause of these severe effects.

The effect of the Maunder Minimum is evident in several historical indices of solar and climatic activity, and is further borne out by studies of the content of carbon-14 in tree-rings (Eddy, 1976). This element is formed in the Earth's atmosphere through a complicated process which varies according to the number of galactic cosmic rays which enter the terrestrial environment.

When solar activity is high, the Sun's magnetic field reduces the number of interstellar cosmic rays that reach the Earth by deflecting them away. Thus, a long interval of low solar activity results in a relatively higher amount of carbon-14 on Earth. John Eddy has documented many of these effects in great detail, and the interested reader is encouraged to obtain these references (Eddy, 1976, 1977, 1983) for a thorough analysis of the topic.

Recent research by Labitzke (1982) and Labitzke and Van Loon (1988) points up another possible connection between the Sun's activity and the Earth's weather. Every two or three years the winds in the lower stratosphere reverse their east to west direction, in an effect which is known as the 'quasi-biennial-oscillation,' or *QBO*. The Labitzke and Van Loon studies show that a relationship may exist between the smoothed solar 10.7 centimeter radio flux and the westerly phase of this oscillation.

No correlation is evident when the flux rate is compared with all stratospheric temperatures at the North Pole. However, a remarkable agreement emerges when temperatures which are associated with the easterly wind flows are removed from consideration. Presently, no one has been able to describe a mechanism which could produce this effect, and this poses a serious obstacle to its acceptance. Unfortunately, stratospheric winds have not been recorded for a sufficient time (data is currently available for less than four sunspot cycles) for a statistically reliable relationship to be demonstrated. In spite of these difficulties, the apparent correlation between these indices is one of the strongest to be proposed.

Further interesting evidence of the Sun's effect upon the Earth's weather has emerged from an analysis of varve samples taken from the ancient glacial lakes of South Australia and other locations (Williams, 1985). These annual sedimentary deposits were apparently laid down nearly 700 million years ago. Each of the layers in the core-specimen is made of material which was carried off in the water from melting glaciers.

The varying thickness of the layers, or varves, is thought to result from small changes in temperature which may be related to the Sun's activity cycle. The record appears to show that the sunspot cycle was substantially the same in ancient times as it is today. The eleven-year period is evident, but the twenty-two-year magnetic cycle is even more apparent. A longer cycle of about 300 years duration is also prominent in the varve samples.

Any relationship between the Earth's weather and the Sun's activity must, by definition, be complex and somewhat tenuous since the effect itself would be so small. It may eventually be learned that solar activity simply 'triggers' a more powerful series of events into play (Giovanelli, 1984). Or it could be that even very small changes in energy, cloud structure or stratospheric winds yield events which are magnified beyond our current expectations. At this time however, the physical mechanism which would initiate changes of a sufficient magnitude to cause an observable change in our weather has escaped the best efforts to detect it.

6

Equipment: the telescope, aperture filters and image projection

There is no need to purchase a particular telescope design in order to make very satisfying, and potentially valuable, solar observations. Consequently, the remarks which follow are mainly intended to be used by those who are new to solar astronomy, or others who are considering the purchase of a telescope which will normally be used for viewing the Sun. **However, all readers must become familiar with the techniques of observing the Sun safely before attempting these observations.** Since these comments are meant to be applied in this manner, they are confined to the types of optical systems which are generally employed by amateur astronomers: the refracting, catadioptric and Newtonian reflecting telescopes.

Common types of telescope

The refractor

The refractor is the oldest of the telescope layouts. Although the optical lens had been in existence for several centuries, the refracting telescope was probably invented sometime between 1590 and 1610 by two eyeglass makers from the Netherlands, Hans Lippershey and Zachariah Jansen (Chambers, 1890). Shortly thereafter, Galileo and his contemporaries refined and enlarged upon the original simple design and popularized its use for studies of the Sun and other astronomical objects. Although these early instruments were very primitive when judged by contemporary standards, they revolutionized the science of astronomy by allowing those who used them to see the stars and planets as never before.

Even now, almost 400 years later, a majority of solar astronomers would undoubtedly recommend the refractor for studies of the Sun, since in most situations it continues to give the best *definition*. Theoretically, this factor increases with the diameter of the objective; and so one would think that the reflecting telescope would produce the best resolution.

However, when viewing the Sun, definition is severely limited by atmospheric distortion, as well as by the degree of accuracy of the objective's figure. Unfortunately, larger diameter objectives are affected disproportionally by

Figure 6.1 Wolf and his successors at the Swiss Federal Observatory employed an eighty-millimeter refractor when determining the daily sunspot number. The official observer (left) and astronomer Thomas A. Cragg are shown in this photograph together with the instrument. Photograph courtesy of Thomas A. Cragg.

adverse seeing conditions. Furthermore, the manufacturing tolerances which are required to achieve the upper limit of a lens's resolving ability are less than those for a primary or secondary mirror (especially with the higher focal ratios). Consquently, it is the refractor which generally yields the greatest image detail.

A second factor guides the choice of instrument for those who plan to study the Sun's spots, and to contribute their data to one of the special programs which compile this information for use by professional astronomers and other researchers. The reason concerns the telescope which Rudolf Wolf and his successors used for many years at the Swiss Federal Observatory in Zurich. This famous instrument is an eighty-millimeter Fraunhofer refractor with a focal length of 110 centimeters (Figure 6.1). When obtaining the daily sunspot estimate, the equipment is used at a magnification of sixty-four, and a polarizing helioscope is utilized so that the Sun's image can be dimmed to a comfortable level (Waldmeier, 1961).

The description of the old Zurich instrument is important information for sunspot observers. Because all sunspot countings since the time of Wolf are adjusted to his scale, it is important that the Zurich instrument be duplicated closely when the observer's estimates are meant to be used in the statistical analysis of relative sunspot numbers. In fact, the observer's estimates can *only* be made compatible with Wolf's values when similar equipment is employed. In this way, the necessary consistency between the values determined by different programs and observers over long time intervals can be maintained.

It is worthwhile to note that under the proper circumstances, an amateur astronomer can easily duplicate an important attribute of the polarizing helioscope. The light which is reflected off the *unsilvered* face of either a narrow-angle prism, such as a Herschel wedge, or a traditional glass prism is at least partially polarized. If an observer has such an instrument, they should obtain a small *polarizing filter* (the type which screws into the eyepiece) from one of the large scientific or optical supply companies, and install it in the eyepiece. The image's brightness can then be varied by simply twisting the entire apparatus slightly in its holder. It is surprising how even small variations in the brightness of the Sun's image will bring out detail which would otherwise be missed. However, polarizers are **not** substitutes for solar filters! **They must be used in conjunction with the normal solar filtering or the observer's eyesight could be severely damaged.**

Because of their generally greater focal ratios, the range of apparent focus is wider (and so, less sensitive) with a refractor than it is for other optical schemes. This feature can prove to be a valuable one for solar observers since observations of fine detail are quite important, and the focus is somewhat different for phenomena which are located near the Sun's limb and those that occur closer to the center of its disk.

However, if a star-diagonal or a similar device which re-directs the optical path to the instrument's side is not present in the system, the refractor's eyepiece location can make viewing awkward when the Sun is at certain angles, particularly when it is near the zenith. In addition, the enclosed tube design of this telescope can occasionally lead to problems caused by a build-up of heat within the tube, with a resulting loss of image definition. This disadvantage can be alleviated to a large extent by turning the instrument away from the Sun every fifteen minutes or so, or by otherwise allowing it to cool; a good practice regardless of the telescope's design. (In order to minimize this problem, the solar telescopes at the National Solar Observatory at Kitt Peak employ cooled and evacuated instrument-tubes.)

For viewing the Sun, the observer should select a refractor with an objective which is at least sixty millimeters in diameter, and a focal length of ten to fifteen times this amount. Today's multiple-component achromatic objective lenses generally do not give rise to the strong secondary spectrum which artificially colored the images formed by the old single-element lenses, so unusually long focal lengths should be avoided.

Although large refractors are difficult to move around because of their long tubes, those with apertures of less than a hundred millimeters or so are reasonably easy to transport. Refractors are well-suited to both projection and direct-viewing methods, and while they are generally more expensive than an equivalent Newtonian reflector, I feel that they still offer the best overall value for the majority of solar observers.

The catadioptric designs

Sophisticated optical systems such as the Schmidt–Cassegrain and the Maksutov offer many of the advantages of the refractor, and (depending upon their size) superior portability. Their design encompasses an optical system which is effectively compressed into a short instrument; a very attractive and convenient alternative to the long, and sometimes cumbersome tube of an equivalently sized refractor.

The superb direct views of the Sun which are obtained with the best examples of this type of instrument are a joy to behold. A number of users also prefer this compact layout for solar photography, where its performance, along with that of the refractor is generally unparalleled. The catadioptric instrument is better suited for use as a multi-purpose telescope than the refractor, because the light-gathering ability of a (generally) larger objective is important when viewing faint objects.

However, the construction of the catadioptric telescope is not as straightforward as it is for the refractor, and some of the materials which are used in its manufacture can be harmed by heat unless the Sun's intensity is diminished *before* the radiation enters the tube. The secondary mirrors of some of these systems are especially likely to be damaged by high temperatures since they have to endure a concentrated beam of sunlight. Those with cemented mountings are particularly vulnerable, and projection should not be attempted with such instruments.

In any event, the manufacturer should be consulted for recommendations before the instrument is used to observe the Sun. Some companies offer special solar filters which are designed specifically for use with their equipment, but may strongly recommend against using the telescope with the projection method.

This design can also suffer from a lesser known problem, in that the catadioptric scheme requires very high standards for its optical and mechanical components if it is to perform at its maximum level. While this may at first appear to be an advantage which would consistently yield an instrument of the highest possible quality, these factors can vary among individual models produced by the same manufacturer. I have found that this is especially true among the older examples of this class, and the result can be an image which is considerably different from the expected view.

Like the conventional reflector, the catadioptric designs are better suited to direct-viewing through aperture filters than they are to the projection method, because the intense heat from the Sun is greatly diminished before it enters the instrument. But the reader must keep in mind that virtually all of the methods for determining the positions of the Sun's phenomena with a high degree of accuracy *require* that projection techniques be used. Consequently, some of these instruments may not be suited to the needs of all observers.

The price of the average catadioptric telescope is usually about that of a good refractor, or occasionally a little higher. However, they are equally well-suited for use in all areas of observational astronomy, and even the smaller-aperture versions will frequently provide excellent views of the Sun. The refractor or the catadioptric layouts are also good choices if the observer has plans for the eventual purchase of one of the new, relatively low-cost hydrogen-alpha filters, which are touched upon later in this chapter.

The Newtonian reflector

The classic reflecting telescope was invented by the Englishman, James Gregory, in 1663. However, it was the illustrious Sir Isaac Newton who modified the Gregorian design into the form which we know so well. Apparently, Newton had long mused over the problems of the refractor, feeling that its lens imperfections could never be overcome. On the other hand, Gregory's invention also had its difficulties, since his system required that a hole be drilled in the center of its mirror so that the emerging rays of light could be examined. In 1669, Newton overcame this complication by installing the secondary mirror at an angle, which re-directed the optical beam out of the side of the instrument (Chambers, 1890).

In the reflector's design, the most serious thermal inequalities originate within the objective (primary mirror) itself. (The infrared portion of the Sun's light can affect the mirror by causing it to expand unequally, and by generating currents within the tube which lessen its resolving ability.) In order to minimize these effects, the mirror should be made of some low-expansion material such as fused-quartz; a feature which is usually (but not always) present in the newer models of reflecting telescope.

Even then, the optical system should be allowed to stabilize for a few minutes before any observations of the Sun are initiated. The focus may have to be adjusted frequently, especially before the optical components have become completely accustomed to their surroundings. It is also important to keep the reflector's tube well covered when it is not in use, since dirty optics absorb and scatter a considerable amount of light.

Almost all reflectors are capable of producing a very good image of the Sun, and the design offers a near-perfect achromatism (although this feature can be ruined by a cheap eyepiece). However, the design is more sensitive to the alignment and collimation of its optical components than the refractor, and is much more likely to need an occasional readjustment of these parts. Newtonian reflecting telescopes can also be difficult to adapt to the projection method because the emerging beam of light exits the instrument from the side, resulting in an awkward location for the projection surface.

On the other hand, the eyepiece is readily accessible from most viewing angles, and this is an important advantage when observing visually. It is hard to over-estimate the benefit offered by a conveniently placed eyepiece; an advantage which the reflector shares with many of the catadioptric systems. High-quality observations and the enjoyment which is gained through the viewing of astronomical objects are dependent to a large degree upon the comfort of the observer, so this feature should not be overlooked.

Ten centimeters is about the minimum aperture for useful solar observations with a reflecting telescope. The higher focal ratios, eight times the aperture's diameter or a little more, are superior to 'wide-field' systems when viewing the Sun, because the size of the primary image is dependent upon the instrument's focal length. Since the optical layout of a reflecting telescope requires that its components be finished and adjusted to a higher standard than those of the refractor, it pays to be certain that both primary and secondary optics are of a similar quality.

In spite of the instruments' potential disadvantages, many successful solar observers use reflectors. They are the least expensive of the popular telescope designs per centimeter of aperture, and this feature along with their light-gathering ability and simple construction has led to their widespread use in all areas of observational astronomy.

No matter what type of instrument is considered, the fewer the number of optical elements, the better. Each component must be of similar quality to achieve the best possible definition in the resulting image, and this can be an unfair (and costly) burden for a mass-produced, multi-element instrument. Also, as the number of optical components rises, the number of times that light must pass between these elements and any thermal currents also grows; a situation which increases the chances for a loss of detail in the image.

The telescope should be soundly constructed and have a substantial (preferably equatorial) mounting. An electronic drive system will certainly prove to be an asset, although it is not needed for normal observations. A mechanical drive *is* a virtual necessity when using the projection method, especially if high-quality drawings or accurate positions of the Sun's phenomena are to be attempted.

Regardless of its type, the instrument should be installed in an area where it can be shaded from direct sunlight. The majority of thermal currents are generated at ground level and so for a temporary installation a grass surface is far better than one of concrete or another heat-retaining material. This is all the more important in the case of the reflector since its primary objective is positioned only a short distance above the ground. Major solar observatories contend with this problem by surrounding the telescope with water (Big Bear Observatory) or by placing their entrance heliostats high above ground.

As was stated at the beginning of this chapter, a good deal of scientifically productive, and very enjoyable viewing can be accomplished with any of the instrument designs which have been discussed. Readers should not miss the wonderfully satisfying experience of viewing the many features of the Sun simply because they feel that they do not have the very best means to observe these phenomena.

By way of example, one successful sunspot observer contributed many thousands of valuable sunspot estimates to the American, Zurich and International programs during his sixty-plus-year career. For better than half a century he used a single instrument: a small Merz refractor with an aperture of only fifty-four millimeters! Thus, in the final analysis, it is the thoroughness, skill and enthusiasm of the observer which yields the best result.

Displaying the Sun's image

Ever since Galileo first turned his telescope to the Sun, astronomers have viewed its fascinating phenomena through a variety of equipment. But at the beginning, the field was fraught with unseen and unknown danger. Consequently, several of Europe's best-known observers suffered permanent damage to their eyesight by using devices which lessened the Sun's brilliance, but not its heat and other invisible radiation. Although Galileo had at least some idea of the inherent dangers of viewing the Sun directly (Drake, 1957), he too was plagued by partial blindness in his later years as a result of his earlier solar observations (Bray and Loughhead, 1965).

During the early development of telescopic astronomy, it was a common practice to view the Sun through blackened, or colored-glass 'filters' mounted at the eyepiece, or by other techniques such as the method pioneered by the famed Sir William Herschel, which employed thinned ink or another opaque liquid as a filtering medium. (The liquids were stored within the eyepiece itself or in glass containers which were placed within the optical system.) Unfortunately, these methods posed a threat to the observer in that they did not fully block the dangerous ultraviolet and near-infrared radiation from the Sun which penetrates the Earth's atmosphere (Chou, 1981).

Several ingenious techniques for reducing the brightness and hidden radiation of the Sun have been devised over the intervening years, and a few have proven to be quite popular with amateur astronomers. Some of the more successful examples include unsilvered mirror schemes and special light-deflecting prisms. However, both techniques must be used with additional filtering at the eyepiece.

Since the early seventeenth century, the Sun has also been studied by examining its projected image. A few authors have ascribed the development of the projection method to Galileo's bitter adversary, Christopher Scheiner. (Along with many others, Scheiner originally believed that sunspots were not a physical part of the Sun, but were small planets revolving about it. His long argument with Galileo over the spots' true nature eventually escalated into a bitter disagreement concerning their discovery.)

However, while a number of engravings which demonstrate the movements of spot groups and faculae as they passed across the Sun's disk are included in Scheiner's great work *Rosa Ursina* (1630), the projection technique was undoubtedly originated by Galileo's greatest pupil, Benedetto Castelli. Furthermore, it is likely that Scheiner heard of the method indirectly, through Galileo's second letter to Welser which was written in response to Scheiner's thoughts on the nature of the Sun's spots (Drake, 1957).

The projection method has been employed by many of the more successful solar observers, including Richard Carrington and Gustav Spörer, and the process remains as the safest of all the ways to view the Sun. When used properly projection provides excellent views of many of the Sun's features, and is recognized as the best method to use when determining their positions.

The modern solar observer generally favors either this technique or that of viewing the Sun directly through a special solar filter known as an 'aperture filter,' which is placed over the front of the instrument. This section presents information on both of these methods, but let me begin with a few suggestions on devices which I believe should be avoided entirely, or used with certain limitations.

First of all, the reader should never use one of the sun-cap (eyepiece) filters which were once commonly supplied with small, inexpensive refractors. They have proven to be **very dangerous**. These gadgets can transmit unsafe amounts of ultraviolet and infrared radiation, or crack without warning, sending a searing flash of sunlight into an unwary observer's eye.

When observing the Sun it is always important to err on the side of safety. Consequently, I recommend that it not be viewed through photographic films, regardless of how they have been prepared. The popular belief that fully exposed and developed black-and-white film can be used as a safe solar filter must be treated with caution. The metallic silver in developed film serves as the filtering medium, and all modern color emulsions (and some black-and-white films) have the silver removed during processing. While these materials may *appear* to be very dark, they are not safe solar filters and, consequently, their use should be avoided.

Welder's glass with a rating of shade number fourteen is satisfactory for non-instrumental viewing, but the material can break if it is used at the eyepiece without some form of pre-filtering. At the same time, welder's glass is not suitable for full-aperture visual or photographic use because its surface is not optically uniform. Shades of less than number twelve or fourteen should not be used *alone* under any circumstances; the weaker shades do not offer the necessary level of protection.

The popularity of the helioscope, or 'solar-eyepiece,' has declined during recent years. Of course this apparatus is really not an eyepiece at all, but instead consists of a special kind of prism, or one or more glass plates mounted within an eyepiece holder. The type of prism which is generally used in this mechanism is known as a *'Herschel wedge;'* a thin, triangular-shaped device which is designed to transmit ninety-five percent or more of the incoming light away from the observer's eye and out of the instrument.

The prism must be mounted properly in order to avoid a damaging build-up of heat within the telescope, and the entire unit **must** be used with a precisely determined amount of supplemental filtering (such as a neutral-density filter) at the eyepiece. Herschel wedge systems have become quite expensive to construct, and are no more satisfactory in use than a cheaper, and safer full-aperture filter.

Viewing the Sun directly with an aperture filter

Aperture filters are generally considered to be a fairly recent innovation, although it is quite possible that their concept dates from an instrument which was constructed in the mid-nineteenth century by the French physicist, Leon Foucault. Foucault frequently examined the Sun's disk with a helioscope whose lens had been covered with a thin film of silver or gold-leaf. When the film was suitably buffed, it reflected away enough of the Sun's light so that the Sun's image could be viewed in comfort (Chambers, 1890).

In my opinion, aperture filters have become the most satisfactory way to observe sunspots, the faculae, and several of the other active solar phenomena. Many experienced solar observers agree that a greater amount of detail is seen when objects are viewed directly, and the aperture filter provides one of the best and safest means to do this. Modern solar filters are generally manufactured of either glass or an optical-quality plastic film, such as *mylar*. These materials are given special metallic coatings which block the harmful radiation from the Sun and reduce its brightness to a tiny fraction of its original amount.

The filters are easy to install by simply slipping the glass filter and mounting-cell onto the telescope's tube, or by carefully securing a mylar filter to the front of the instrument. In this way, it is an easy matter to convert the telescope from night-time use into the unique world of daylight astronomy.

Unfortunately, direct viewing does not easily lend itself to the task of deriving exact positions for the Sun's phenomena. Of course, if the Sun's disk is to be photographed (rather than drawn or traced) as a basis for these reductions, the aperture filter can be used with good success. More often than not, however, the photographic image of a spot cluster is not distinct enough to be measured with a high degree of accuracy when it is obtained with the level of equipment which is available to an amateur.

Aperture filters which are made of glass often produce a yellow—orange image of the Sun. This will prove to be an advantage for many observers since the eye is more sensitive to fine detail when it is viewed in this portion of the spectrum than it is at the shorter wavelengths. (Mylar yields a bluish image of the Sun because the plastic material blocks slightly more of the near-infrared radiation than glass does.) Glass is also more durable than mylar, and an extra-hard metallic coating, usually a chromium alloy such as Inconel, can be applied to it.

On the other hand, glass filters are susceptible to outright breakage if they are mishandled. And, while a mylar filter can be attached to the instrument with less-formal methods, the glass filter requires a special mounting. Be aware though, that if the filter is installed too tightly in its cell it could crack suddenly during use. To minimize this (somewhat rare) possibility, place a new filter within its mounting cell outside in the Sun and allow the combined unit to warm for a *few minutes* before attempting to use it for the first time. **Warning!** Do not overdo this safety test. Heat is *accumulated* during a continued exposure to the Sun, and if the unit is left for more than fifteen minutes or so the resulting expansion could crack a filter which is too-tightly mounted.

Full-aperture filters are fairly inexpensive, easy to store, and are available in sizes that fit virtually any size of instrument. When not in use, the filter should be stored in its mounting inside a box or other protective device. While they are generally a bit more expensive than their mylar counterparts, glass filters still represent an excellent value and their benefits should not be under-estimated.

Filters which are manufactured of mylar or a similar space-age material are the least-expensive type of solar filter, consistent with safety and convenience. The aluminized mylar coating provides a safe filtering medium by reflecting away some of the Sun's radiation, and by absorbing additional amounts. In this way, the total quantity of transmitted-radiation is reduced to only about *one-tenth or one-hundredth of one percent* of its original amount.

A few of the Sun's brighter phenomena, such as the rare, solar 'white-light' flares and the faculae, may be slightly easier to detect when viewed in light which is shifted towards the blue region of the spectrum, as it is with a mylar filter. This comes about because a few of the Sun's features emit a large portion of their total energy at the shorter wavelengths, and consequently they may appear to be a bit brighter when they are viewed nearer to this spectral region. On the other hand, a number of observers feel that the blue image which mylar offers is less pleasing to the eye when viewing the Sun's spots than the more familiar color which is transmitted by a glass filter.

One point is often overlooked when mounting the mylar material onto the instrument. This light-weight material **must be securely attached** to the telescope, or otherwise the filters can be blown off by a sudden gust of wind. This may seem to the reader to be quite obvious, but it is surprising how frequently this precaution is ignored by observers who are very safety conscious in other ways and should know better.

Mylar filters are probably more prone to pinholes in their coatings than glass filters are, because they are quite flexible and their coatings are usually not made of the hard alloys which can be applied to glass. If these defects do occur, and are allowed to go unchecked, harmful amounts of radiation can eventually enter the instrument. The application of aluminized coatings to both sides of the plastic film has reduced this problem considerably, but not entirely.

A filter's coating can also become damaged through abrasion or misuse, and so it should be monitored on a routine basis. To check for the presence of these defects, hold the filter up to a strong light and inspect it for thin areas and pinholes in its coating *before* its initial use, and then at regular intervals thereafter. A few tiny holes or a small abrasion can safely be repaired with a small dab of black paint, but any filter with more serious problems should be returned to the manufacturer.

The performance of either type of filter, especially when it is used with a Newtonian reflector or a catadioptric instrument, can occasionally be improved by reducing the telescope's effective aperture. It is a good idea to experiment with 'aperture-stops' of different sizes if the instrument has an objective diameter which is larger than fifteen centimeters for a reflector, or ten centimeters for a refractor.

The diaphragm should be placed over the existing aperture, with the new opening centered *between* the secondary optic and the tube-wall for reflecting systems, and centered in front of the objective lens in the case of a refractor. The maximum size of the new aperture is often dictated by this space of course, but otherwise a good starting point for the diameter of the reduced opening is around seventy-five millimeters. Stops may be constructed of any smooth and light-weight material such as a good-quality cardboard, and should be painted dull-white.

When contemplating the use of a diaphragm, it is important to remember that an instrument's resolving ability is almost always effectively less in the daytime than it is at night. The difference is due to the normal atmospheric instability which is caused by the Sun's heat. The small light-loss which results from a reduced aperture is not important when viewing an object as bright as the Sun, and because slightly less heat is introduced into the optical system when the aperture is decreased in this manner, definition may actually be improved.

All aperture filters offer an important and unique advantage in that they block much of the Sun's heat *before* it can enter the telescope. They are suitable for use with all types of instruments, but this feature makes them a particularly appropriate choice for use with catadioptric instruments and Newtonian reflectors. As we have discussed, the most serious difficulties that are encountered while observing the Sun arise when unfiltered sunlight is allowed to enter the telescope.

Full-aperture filters provide an excellent way to view the Sun. Always purchase them from reputable companies (the telescope manufacturer itself is often a good choice), examine them carefully before and during their use, and treat them as valuable optical accessories. In turn, they will reward you with many hours of safe and enjoyable viewing.

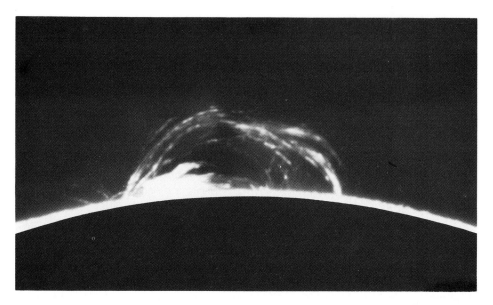

Figure 6.2 A loop prominence is shown in this hydrogen-alpha (Hα) photograph supplied by Jean Dragesco. Although Dr Dragesco employed a narrow-band (sub-ångström) Hα filter for this photograph, prominences and other limb phenomena can also be viewed with the less-expensive, wider-band Hα filters.

Hydrogen-alpha filters

Until recently, the majority of amateur astronomers could never hope to view the Sun in other than white-light. However, new developments in technology have resulted in the entry into the marketplace of a relatively inexpensive apparatus known as a '*hydrogen-alpha filter.*' A hydrogen-alpha (Hα) filter is a special type of device which transmits only a specific wavelength of light, usually centered around 6563 ångströms in the red portion of the spectrum. By isolating the light of hydrogen emission, Hα filters permit views of many of the most spectacular solar phenomena, such as the *prominences* (Figure 6.2).

Prominences are gigantic eruptions of dense hydrogen, which can be viewed extending above the Sun's surface to heights of 80,000 kilometers or more. When they are seen projecting from the Sun's limb they are called prominences, but when they are viewed against the disk background they appear as dark sinuous lines and are known as *filaments*.

All Hα filters are capable of showing limb prominences, but only narrow-band filters (those with band-widths of less than one ångström) allow features of the Sun's disk to be seen. When they are viewed in this manner solar flares appear as intensely bright areas, while active sunspot regions show a wealth of fine structure which is not available to a conventional observer. Several of these phenomena can be discerned by examining Figure 6.3. The great differences in

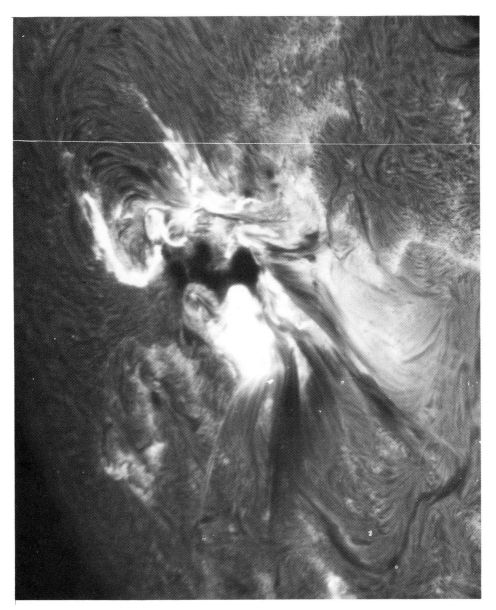

Figure 6.3 A spectacular solar flare *occurs in a complex sunspot group during March 1989. Note the filamentary structure and other solar features which are shown in this superb photograph supplied by Thomas G. Compton.*

the Sun's appearance when it is viewed in the Hα band, and again in white-light, are clearly demonstrated by the photographs of a large sunspot group taken only moments apart by Dr Jean Dragesco from the south of France (Figure 6.4).

Before purchasing any Hα filter, however, the reader should inquire about its

65

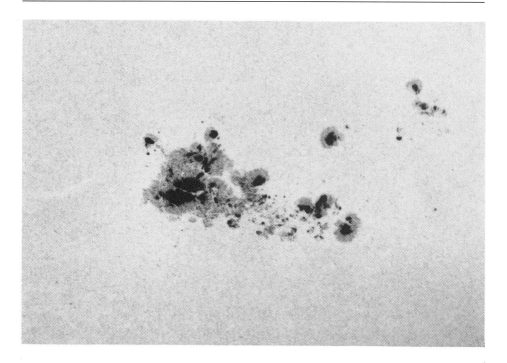

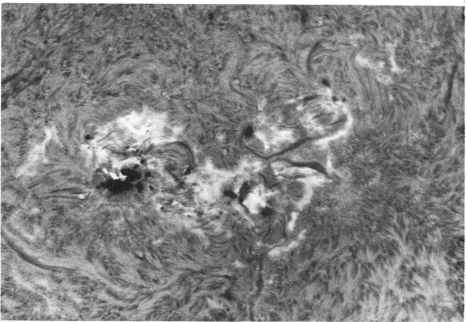

Figure 6.4 In order to observe features on the Sun's disk, a Hα filter with a narrow passband width must be employed. Note the great difference in appearance between these photographs of a large spot group taken less than one hour apart in white-light (top), and in Hα. Photograph courtesy of Jean Dragesco.

features and limitations with respect to the telescope with which it will be used, and should be certain that it will meet his or her needs as an observer. A filter's cost generally goes up as the width of the bandpass decreases, but it is the sub-ångström (narrow-band) filters which show the greatest detail. Consequently, these may represent the best value over time, when their owner's thirst for a more extensive view of the Sun's phenomena may have outgrown his or her initial curiosity.

Projecting the Sun's image

Even though aperture filters have been shown to be very safe devices, the technique of projection is still the safest of all ways to observe the Sun. Although most observers feel that detail on the Sun is best seen when viewed directly, the projection method also has its strong advocates. Eyestrain is a good deal less with projection than it is when viewing the Sun directly, and the method offers a further advantage in that it allows several people to view the image simultaneously.

As we have mentioned previously, projection is the method-of-choice for those who wish to determine the heliographic positions of sunspot groups or other of the Sun's phenomena. The technique is particularly well-suited for use with refracting telescopes. Due to the location of the eyepiece, projection is a little more difficult to apply to a Newtonian reflector; however, it is often successfully employed with these telescopes as well. An electrically driven, equatorial mounting is not absolutely necessary, but it is highly recommended for use with this method.

On the negative side, projection is seldom satisfactory when it is used with refractors which have less than sixty millimeters of aperture, or with reflectors of less than twice this size. Again, it is important that owners of catadioptric telescopes consult the manufacturer of their instrument **before** using this method, in order to avoid subjecting their equipment to possible damage.

Regardless of the instrument's optical layout, the eyepiece used to project the image should be chosen with care. Those with cemented components or anti-reflection coatings should not be used because the Sun's heat is concentrated in the eyepiece area and it can soften these materials. The older Ramsden or Huygenian types should have their lens-retaining rings loosened slightly to allow for expansion. (A slight 'rattle' should be heard when the eyepiece is gently shaken.)

The Sun's image should be projected onto a surface which is mounted perpendicular to the emerging cone of light. The little projection apparatus which is included with some of the smaller instruments is not suitable for serious observations. The surfaces of these accessories are invariably too small, and their flimsy mountings allow the screen to vibrate excessively.

The surface itself should be fifteen to twenty percent larger than the whole disk image so that the entire Sun can be displayed at one time. The screen should be constructed of a smooth and flat material such as hardboard, and painted dull-white. Some provision should be made so that the distance between the eyepiece and surface can be adjusted if a constant image diameter is required, since the Sun's angular diameter varies slightly as the Sun-to-Earth distance changes throughout the year.

The diameter of the projected image is an important consideration. This aspect is somewhat dependent upon the instrument's size, but I would suggest that a diameter of fifteen centimeters is a good choice for most amateur applications. This size has come to be recognized as a sort of 'international standard' for amateur solar astronomers. Further, it will allow the use of the Porter Disk, a simple device which is useful for determining the accurate position of a spot group.

The image's diameter can be controlled in two ways: through changes in the eyepiece magnification, and by varying the distance between the projection screen and eyepiece. The relationship between these aspects while the Earth is at its mean distance from the Sun is defined according to

$$\text{distance between surface and eyepiece} = \frac{107D}{M-1}$$

In this equation, M represents the magnification of the eyepiece, and D is the required image diameter in centimeters. The magnification for a given eyepiece varies according to the instrument's focal length and is obtained by dividing the focal length of the telescope by that of the eyepiece. For example, an instrument with a focal length of ninety centimeters would require an eyepiece with a focal length of fifteen millimeters in order to produce a magnification of sixty.

It follows that the desired diameter of fifteen centimeters can be obtained with an eyepiece which magnifies the image by a factor of sixty and an eyepiece-to-projection surface distance of around twenty-seven centimeters. Since the distance between the Sun and Earth changes continually, this measurement will need to be varied by about four millimeters if the same diameter image is to be maintained throughout the year.

The most satisfactory location for the projection screen is in the *darkest location possible*. Depending upon the resources (and the resourcefulness) of the observer, the Sun's image might be projected into a simple projection box, onto a screen attached to a mounting stand which may or may not be attached to the instrument, or even into a darkened room.

If at all possible the projection surface should not be physically attached to the telescope. Several observers have had success with mounting the screen on an old music stand or a similar object. The stand should be one of the older professional styles which are adjustable in height and angle and are quite heavy, as should any mounting which is separated from the instrument.

On the other hand, if it is necessary to attach the projection surface onto the telescope, every attempt should be made to make it as light-weight and free from vibration as possible. In order to shield the image from direct sunlight, a baffle which blocks the direct rays of the Sun should be constructed, or the image can be projected into a box which is secured to the instrument.

When an attached, but otherwise conventional projection screen is chosen, the baffle which shields the image can be constructed of heavy card, thin sheet-metal or a similar material. It should be mounted so that it will completely shade the projection surface. The shield's placement and size varies with the instrument's type and aperture, and can easily be found by trial and error.

On the other hand, if the image is projected into a box, the enclosure's interior walls should be painted dull-white (not black, which absorbs heat), and an opening provided so that the Sun's likeness can be viewed with ease. The box should be affixed in such a way as to be adjustable in its distance from the eyepiece and also so that it can be rotated about the optical axis. The latter adjustment will need to be varied considerably (by up to fifty-three degrees) during the orientation of the Sun's image, a procedure which is explained in the following chapter.

When the instrument that is used to project the image is a fairly large one, or if the projection screen is to be mounted separate from it, a rotating projection surface might be tried. Several observers have constructed rotating screens and I can attest to the effectiveness of this technique when it is compared with the more typical means of projection. The rate of rotation for such a device should be between ten and twenty revolutions per minute. It can be turned by hand, although a small electric motor of the type generally used with old phonographs is a better choice.

The rotating screen minimizes any defects in its surface and allows a larger proportion of the image's detail to be seen. The slow speed of rotation should not adversely affect the view, but any excessive vibration must be avoided. I leave the actual construction of this device to the ingenuity of the reader, but when the apparatus is properly designed and constructed, an extraordinary view of the Sun can be obtained.

Image orientation

Whenever the features of an object with a large angular diameter such as the Sun are observed or studied, it is important that the correct cardinal directions be assigned to the image. Since the Sun's appearance varies widely according to a number of factors, this topic is often a confusing one for those who are new to observational astronomy. However, the same general orientation procedure is applied in all situations, and it is in fact, a relatively simple process.

Congruent and incongruent images

One factor which affects the appearance of an image is its congruency: that is, how the image which we see coincides with the object itself. Thus, a 'congruent' image is what one sees by eye. Whether an image will be congruent or not is pre-determined by the optical configuration of the instrument which is used to view the subject, and by the technique which is employed to display it. The following experiments were designed by David Rosebrugh in 1986 in an effort to help clarify these matters. (If the reader is confident in his or her understanding of the effects of reflections and other conditions which affect the appearance of an object when it is viewed with optical aid, this section can be overlooked.)

Example one

Look with the unaided eye at a house at some distance. The image which you see is a congruent image. For example, the house is upright and the garage is located on its *right-hand* side. Now look at the house through a simple finder-telescope; the type without a prism or other reflecting device which would redirect the optical path. The house and garage now seem to be upside-down. The image is still congruent however, because the garage remains in the same location relative to the house. Imagine yourself standing on your head while looking through the finder. The house will be magnified somewhat of course, but otherwise it will appear exactly as it did when it was viewed without optical assistance.

Now insert a small prism or similar right-angle viewing device into the finder along with a low-magnification eyepiece, and look at the house again. Now the image is 'incongruent.' That is, the house continues to appear upside-down, but the garage now seems to be located on its *left-hand* side. The congruent image which emerged from the basic lens system has been flipped once by the reflection from the prism, forcing it to be incongruent.

Finally, look at the house through one barrel of a small (prism-type) binocular. In principle, this is nothing but a small finder-telescope in which the upside-down congruent image has been made upright by specially arranged reflecting prisms which 'flip' the image four times. Each flip, or reflection, reverses the image once, so that four flips return it to being congruent, and cause it to appear upright. (In the binocular, the last of these reflections also brings the image out to the eye.)

Example two

Obtain a sheet of clear plastic film with printing on it, such as an old bread-wrapper. Have a friend hold it out with arms extended while facing you so that you can read the printing. Of course, the image which you see is a congruent image. On the other hand, your friend behind the wrapper sees an image which is analogous to one which is incongruent because left and right appear to be reversed.

Your friend can rotate the plastic so that the top is at the bottom or the side, but no matter the angle, the image continues to be incongruent. In effect, the printing has been reversed. This procedure is similar to one in which the observer looks downstream (in the direction of travel of the Sun's rays) at an image which is *projected* rather than viewed directly; a situation which also results in an apparent flip.

Thus in this sense each flip is actually a reversal from side to side. One reflection causes a single flip, and two reflections result in two reversals. If the image is congruent before encountering these circumstances, an even number of reflections will restore it to congruency. A traditional Newtonian reflecting telescope produces a congruent image (two reflections: one from the primary mirror and one from the secondary mirror). On the other hand, a typical refractor with a prism such as a Herschel wedge or star diagonal yields an incongruent image because there is only a single reflection.

The effects of time and the observer's position on the image

Any observer, without regard for his or her location or equipment, will see the Sun's disk rotated to many angles when it is viewed telescopically. The angle will change significantly depending upon where the observer stands at the eyepiece: at the telescope's left, behind it, or on its right. And, the disk's angle will also appear to change as the day progresses from morning to evening. Observers who use dissimilar types of instruments may encounter slightly dif-

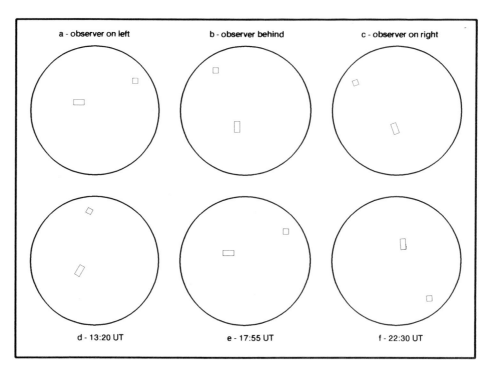

Figure 7.1 The appearance of the solar disk changes according to the observer's position relative to the eyepiece, and as the day progresses.

ferent consequences, but no matter what type of equipment is employed, the angle will vary according to these two circumstances.

Similarly, observers south of the Tropic of Capricorn should be aware that the Sun appears to be noticeably rotated when compared with its appearance at an equivalent northern latitude and time. On the other hand, those who live in the Tropics see the Sun exactly as do Northern Hemisphere observers north of the Tropic of Cancer during part of the year. But at other times, when it is north of them, they see the Sun as do those who live south of the Tropic of Capricorn. The apparent motion of the Sun along the horizon is reversed for those who live in the Southern Hemisphere, moving from *right to left* during the day when it is viewed looking northward. (The path of stars as they rotate around the South Pole of the sky is also opposite to the Northern Hemisphere view; they appear to move *clockwise* rather than *counter-clockwise*.)

Figure 7.1 shows the Northern Hemisphere view through a refractor with a Herschel wedge prism, so another observer's view may not match these sketches exactly (the drawings may be reversed, etc.). For the remainder of this chapter my remarks are intended specifically for readers who live in the Northern Hemisphere. However, the same principles will apply regardless of the observer's location.

The six sketches which illustrate these changes, Figures 7.1a to 7.1f, were made on the same day with the eyepiece axis as near to the vertical as possible. Since a prism was present in the system, the cone of light emerged at a right-angle to the refractor's normal optical axis when the drawings were prepared. Figures 7.1a, 7.1b and 7.1c were constructed around 13:30 Universal Time (UT). Sketch 7.1a was made while the observer was standing on the instrument's *left-hand* side, sketch 7.1b when directly *behind* the telescope, and sketch 7.1c while on its *right-hand* side.

In effect, the observer moved in a counter-clockwise direction around the eyepiece, and this movement caused the image of the Sun to appear to rotate. Thus it is obvious that the appearance of the Sun can change a great deal depending only upon where one stands in relation to the telescope. The same is true when projection is used. The disk's orientation depends upon whether the image is projected to the left, to the right or directly behind the telescope.

Figures 7.1d, 7.1e and 7.1f demonstrate the varying angle of the disk during a typical day. In order to neutralize the effects of the observer's position relative to the telescope, these sketches were all prepared with the astronomer standing directly behind the instrument, looking downward into the eyepiece. Sketch 7.1d was made at 13:20 UT, sketch 7.1e at 17:55 UT, and sketch 7.1f at 22:30 UT. Thus, in a little over nine hours, the Sun again changed its appearance dramatically. Since these latter variations occur continually throughout the day it is obvious that one cannot simply guess at the orientation of the Sun!

None of these circumstances influence the following methods of determining the cardinal directions on an image or a sketch. These techniques simply 'standardize' a drawing without regard for the instrumentation or method which is used to view the image. In each situation we will apply the information to a simple sketch of the solar disk, a highly recommended procedure which will prove to be useful in understanding the continuously changing face of the Sun.

Determining the directions on an image

Whether we project the Sun's image or view it directly, it is a very easy matter to assign rough directions to its disk. We need only perform two simple tasks to do this. First, because of the Earth's rotation the Sun's image will always appear to move through the field in a direction which is from the east and towards the west. Consequently, if we stop the telescope's drive motor and allow the image to 'drift' through the field, the Sun's western limb will disappear first. To determine north and south, we move the front of the instrument towards north in the sky while simultaneously viewing the movement of the image. The southern limb will be the first to disappear.

The determination of these directions brings to light an anomaly which occurs when defining east and west on a drawing of the Sun. The western limb is considered to be its leading limb; just the opposite of the west-to-east rotation of the Earth. We designate the limbs in this manner because directions on the Sun are assigned according to the way we view them. In fact, both the Earth and Sun rotate counter-clockwise as seen from above their North Poles.

Although the rough directions which are obtained by these two steps are interesting and somewhat informative in themselves, they will seldom meet the needs of a serious observer. Consequently, in the succeeding sections I will describe an extension to this procedure which will result in an exact orientation of the Sun's image.

Projected image

Prepare a drawing of a circle with a diameter which is exactly the size of the projected image, and project the Sun's image upon it after the manner which has been described in the previous chapter. Now, choose a sunspot somewhere near the disk's center and mark its position on the drawing, labeling it 'a.' Then turn off the telescope's mechanical driving system, and allow the image to drift through the field of view.

In a few moments, spot 'a' will have moved to 'a'.' Mark this second position on the drawing as well. The line drawn between these two points which is shown in Figure 7.2a, represents the actual east to west drift-line in the sky. Now construct a second line (E–W) parallel to the drift-line which passes through the center, or 'origin', of the circle. Since the Sun appeared to drift *towards* a', 'W' represents the western limb of the Sun.

Finally, draw a third line perpendicular to the E–W line which also extends through the origin. This line, labeled N–S on Figure 7.2b, represents the north–south line in the sky for the Sun's position at the time of the observation; it does *not* depict the Sun's rotational axis which will be determined later on. Its northern end will be indicated by the simple procedure which was described at the beginning of this section. Mark these directions on the sketch to complete the preliminary orientation of the projected image.

Directly viewed image

The process of locating the directions on an image which is viewed directly is very nearly the same as that described for one which is projected. However, in this situation the Sun's likeness will usually appear to be reversed by 180 degrees when it is compared with the projected image.

An eyepiece which allows the entire disk of the Sun to be seen at one time should be chosen for this procedure. In addition, the process will be much simpler if the eyepiece is equipped with a 'crossed-line' reticle, which can be obtained from any large scientific or optical supply company. The eyepiece which is selected must be of the 'positive' type, such as a Kellner design, or the crossed lines cannot be brought to a focus. The directions can be properly placed

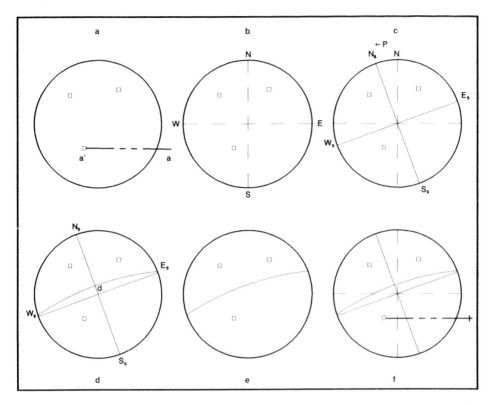

Figure 7.2 These diagrams illustrate how one determines the precise directions on the Sun's disk and correctly locates the solar equator according to a series of steps which are described in the text.

without this helpful accessory, but I strongly recommend that the reader purchase one of these devices, especially if he or she intends to orient the Sun's image according to the variations which occur each day (see below).

At any rate, for the remainder of this chapter I will assume that the reader is using a crossed-line reticle whenever references are made to images which have been viewed directly. The observer should also prepare a drawing of a circle (perhaps sixty millimeters in diameter) which is bisected by two perpendicular lines (see Figure 7.3).

To determine the east and west directions in the sky, rotate the eyepiece/reticle unit so that a sunspot moves parallel to one of the crossed lines as the Sun drifts through the field of view. Label east and west appropriately on one of the bisecting lines, remembering that, as always, the direction of travel is from the east and towards the west. Mark the north and south directions on the other line, after determining them as explained previously.

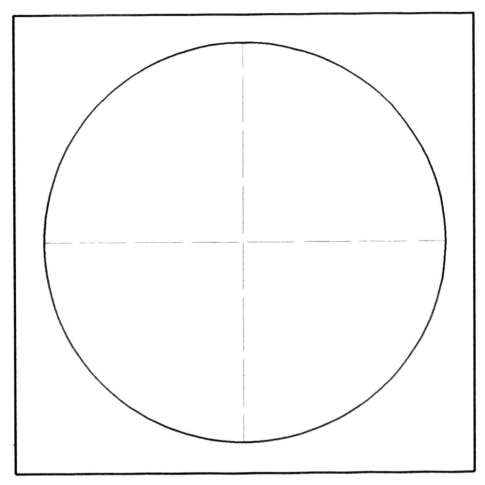

Figure 7.3 When deriving the positions of the Sun's spots or other features, the phenomena should be sketched onto a basic form such as that shown here.

Precise orientation of the image

Before beginning, secure a table of the Sun's daily orientation angles for the components, P (the position angle of the Sun's true axis of rotation) and B_0 (the tilt of the Sun's North Pole relative to the Earth) from a source such as the *Astronomical Almanac*. A more extensive description of these quantities is included in Chapter 9, but typical examples of the changes in the positions of the Sun's axis and equator during the year are shown in Figure 7.4.

To better illustrate the following process it has been sub-divided into four steps, each of which is depicted in Figures 7.2c–f. These steps continue the procedure which began with the determination of the directions on the image. After the technique is understood, the reader will presumably proceed directly to the single sketch shown in Figure 7.2f, with all steps depicted on the one

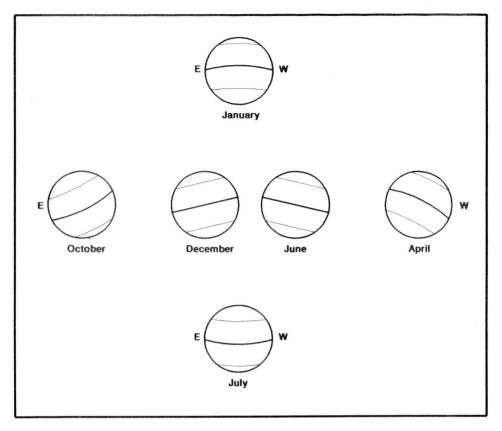

Figure 7.4 The appearance of the solar disk undergoes large changes during the year which are caused by the continuously changing orbital geometry between the Sun and Earth. The apparent movement of the Sun's axis and equator relative to an observer on Earth is illustrated in this diagram.

diagram. For our example, let us assume a date of 1 February for the observations, and a value of $-12.0°$ for P. A negative value for P indicates that its angle should be measured towards the west from the north-point, N, on the drawing. Positive angles are measured towards the east.

Measure and mark this angle on the sketch, and construct a line (N_s–S_s) which extends from this point through the disk's center to the opposite edge of the circle. Now draw a second line, (W_s–E_s), perpendicular to N_s–S_s, which also passes through the origin. These steps are illustrated in Figure 7.2c. The points W_s and E_s symbolize the places where the Sun's equator intersects the edge of the disk, and the N_s–S_s line represents the Sun's actual axis of rotation.

Next, select the value of B_0 for the date of the observation from the table in the *Astronomical Almanac*. Let us assume that it is $-6.0°$. Negative values indicate that the Sun's North Pole is tilted *away* from the Earth by the amount of B_0, and positive angles show the opposite effect. Therefore, in this case the solar equator

77

will pass across the N_s–S_s line to the *north* of the drawing's center. The distance, d, from the center of the circle to this intersection is calculated according to

$$d = R \sin B_0,$$

where R is the radius of the circle.

Since the sine of six degrees is 0.1045 and the radius of the circle in Figure 7.2d is assumed to be thirty millimeters, 'd' will equal a bit over three millimeters in this situation. Plot this point on the N_s–S_s line. The Sun's equator can then be represented by an ellipse which is drawn through the points W_s, d and E_s (Figure 7.2d). Note that B_0 always decreases to zero at the limb.

Figure 7.2e shows only the Sun's equator, spots and axis of rotation; all other construction lines have been eliminated. After examining the drawing we can see that there are two Northern Hemisphere sunspot groups and one southern cluster.

Figure 7.2f is a composite diagram showing all of the steps which have been described. The image of the Sun which is illustrated by these figures is incongruent. A congruent image would be labeled, moving clockwise from point N_s; W_s, S_s and E_s. This is exactly as the Sun appears in the sky when it is viewed without optical assistance at midday.

Determining a sunspot's position mathematically

The heliographic position of a spot group can be determined mathematically from a sketch which has been constructed while viewing the Sun directly. However, the accuracy of the result will seldom equal that which is obtained with a graphical technique such as the Porter Disk method because the spot's position must be estimated, rather than traced or photographed.

If you would like to try this method, any sunspot groups or other features should be lightly sketched onto the drawing before beginning the orientation procedure. Use the crossed lines on the reticle and the limb of the Sun as reference points for drawing on the groups, and locate them as carefully as possible. It is neither necessary or desirable to add detail to the sketched groups; instead their positions should be indicated by small rectangles or circles. The degree of care which is exercised at this stage of the process is the single most important part of this technique.

After the sketch has been completed, proceed as outlined above to the result shown in Figure 7.2f. Then complete the following process (Roth, 1975) which will be considerably easier if a scientific calculator is used to solve the necessary equations.

First measure the distance from the center of the circle to the spot in question. When making this measurement, choose the mid-point of the main preceding spot of a bipolar group as a reference point because it will frequently be the longest-lasting member of the cluster. Otherwise, measure to the center of a unipolar spot or other feature. Let us say the result is twenty millimeters. Call this component r. Now compute ϱ according to

$$\sin \varrho = r/R,$$

with R again set equal to the radius of the circle, or thirty millimeters.
Thus,

$$\sin \varrho = 20/30 = 0.6667 \quad \text{and} \quad \varrho = 41.81 \text{ degrees.}$$

Now calculate b, the heliographic latitude of the spot:

$$\sin b = \cos \varrho \sin B_0 + \sin \varrho \cos B_0 \cos \varphi.$$

The quantity φ is the measured angle between the North Pole of the Sun's axis of rotation (N_S on Figure 7.2f) and a line drawn from the origin through the

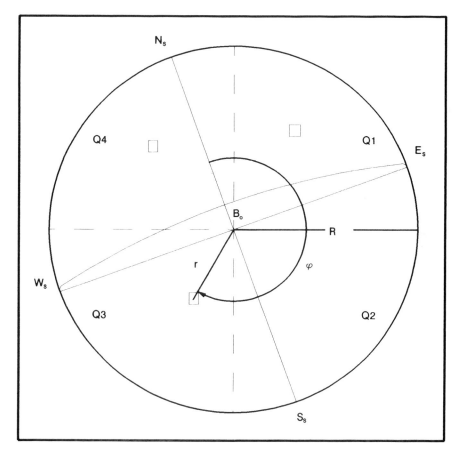

Figure 7.5 Although the method is not as accurate as those which utilize a projected (or photographed) image, the position of a spot group can be determined mathematically from a drawing made at the eyepiece.

Table 7.1

Quadrant	sin	cos
1st	+	+
2nd	+	−
3rd	−	−
4th	−	+

center of the spot. In the case of our example the angle is 210 degrees (Figure 7.5). The sign of φ varies according to the quadrant where the spot lies and the trigonometric function. These relationships are listed in Table 7.1.

So again assuming a value for B_0 of −6.0 degrees, and solving

$$\sin b = (0.7454)\,(-0.1045) + (0.6667)\,(0.9945)\,(-0.8660)$$
$$= -0.0779 + (-0.5742) = -0.6521;$$

we find that

$$b = -40.70 \text{ degrees.}$$

The heliographic latitude in this example is negative. Consequently, the spot is located in the Sun's Southern Hemisphere.

To derive the cluster's longitude, first calculate l, the distance of the spot from the Sun's central meridian:

$$\sin l = \sin \varphi \, \sin \varrho \, / \cos b.$$

Thus,

$$\sin l = (-0.5000)\,(0.6667) \, / \, 0.7581$$
$$= -0.3334 \, / \, 0.7581 = -0.4398.$$

In this case, l amounts to −26.10 degrees. From the table in the *Astronomical Almanac* which we have used previously, we find that the longitude of the central meridian, L_0, for zero hours Universal Time on the date of the observation, is 148.0 degrees. However, the tabular value should be corrected through the process of linear interpolation so that it reflects the actual time of the observation before it is entered into the final computation. While performing this procedure, note that L_0 *decreases* by approximately 13.2 degrees per day throughout each of the Sun's rotations. Consequently, if the observation was made at 18:00 Universal Time, L_0 would have a value of 148.0 degrees less (13.2 degrees multiplied by 18/24), or 138.1 degrees.

Ignoring the consequences of any required interpolation, the heliographic longitude of the group of spots will be given by

$$L_0 - l:$$

and so,

$$148.0 - (-26.10) = 174.1 \text{ degrees.}$$

Since in this situation the calculated longitude is greater than the tabular value, we know that the spot is located to the west of the central meridian.

Again, the heliographic positions which are obtained in this manner are not normally as precise as those which are derived through the use of graphical techniques. However, they are sufficiently accurate for many applications including those which require the assignment of correct hemispherical locations or for other purposes that do not require a high degree of accuracy.

8

Observing the Sun's spots and other solar phenomena

Caution: Readers should not attempt these observations until they have read and fully understood the portions of this book which cover instrument design and observing methods. **Serious eye damage can result if the simple precautions that are explained in these sections are not followed.**

Very few areas of observational astronomy offer the challenge and potential which the Sun presents. Apart from that, it's a lot of *fun* to observe the Sun, and especially those enigmatic phenomena, the sunspots. Sunspot numbers increase in importance with each passing year, because they reach back further in time than any other index of solar activity. This continuous daily record serves as an important link with the past for studies of the Sun's long-term variability. The regular measurement of sunspot numbers by dedicated amateur astronomers does not require sophisticated or expensive equipment, and provides one of the few remaining opportunities for an amateur to contribute meaningful data to the professional scientific community.

Other than the daily counting of the spots and groups, the assiduous observer should be alert for the following: *solar white-light flares* (the observation of these rare events is discussed in Chapter 10); *faculae* and *light-bridges*; *veiled spots* and *wisps*; *color* within sunspot groups; and the *growth, change and movement* of particular groups, as well as their size.

The system for dividing sunspots into groups and counting the spot and group combinations has been covered in previous chapters. In this section I will make several suggestions which are meant to aid the development of the observer's technique, and call attention to some of the interesting solar phenomena that may be encountered during a typical observing session.

Our first duty is to bring the Sun's image into the telescope's field of view. Before beginning the observing session, securely cap or otherwise cover the instrument's finder-telescope. It is not required for solar observations, and this precaution will protect an unwary bystander from possible harm. Now remove the full-aperture filter (if the observations are to be made directly) and the eyepiece.

Point the instrument towards the Sun and adjust it so that the shadow of the tube is at its smallest. Hold your hand or a small card a few inches behind the eyepiece holder and adjust the telescope slightly so that the bright solar image appears. Follow the same procedure when using the projection method by bringing the image onto the projection screen. Replace the filter and eyepiece and make any final adjustments which are necessary to center the image in the field.

Next adjust the focus so that the image of the solar limb is sharply defined. If sunspots are present, the focus should be fine-tuned on them so that the smaller spots are as distinct as possible. If, as is common at sunspot minimum, no spots are immediately evident, try to fine-focus on the faculae so they appear at their brightest, or on the solar granulation if the observing conditions will allow it. This procedure should also be followed if projection is chosen as the viewing technique.

Although a detailed drawing is not necessary when counting sunspots, an informal sketch of the solar disk should be made each day. When preparing the drawing show the approximate location of each spot-cluster by noting it on the sketch as a small circle or square. Record the number of individual spots within each group directly on the diagram in a manner similar to that which is shown in Figure 8.1.

Now determine the approximate cardinal directions according to the methods previously described. When the Sun is observed at nearly the same time each day this step need be done only once; at the time of the initial observation. If this procedure is followed faithfully, each sketch will serve as a guide to the following day's activity and any faint groups can be more easily located on subsequent days. This process is far more interesting and informative if the solar equator is sketched onto the drawing, and the groups are assigned to the correct hemispheres. Of course if the latter procedure is followed, each day's sketch will need to be similarly refined so that changes in the geometry of the Sun's disk are evident.

All astronomical observations should be recorded according to a common time-frame. For studies of the Sun, the time system which is usually employed is *Universal Time*. Universal Time (UT) is recorded from 00:00 hours to (but not including) 24:00 hours, which is recorded as zero hours on the following day. Each day officially begins at 00:00 UT at Greenwich, England. Table 8.1 lists the necessary corrections for various standard time zones in the Americas (Mayall and Mayall, 1968).

The appropriate tabular correction is added to the observer's local time. If the local time is 1pm or later, add twelve hours to conform to the twenty-four-hour system. If the result is greater than twenty-four hours, subtract this amount from the total and record the difference as having occurred on the following day. Pay particular attention to this change of date if it is mandated by the difference between local time and Universal Time. If daylight-time is in effect, *subtract* one hour from each correction value before conversion. (For instance, the correction for the Eastern Standard Time zone during intervals of Daylight Savings Time is four rather than five hours.)

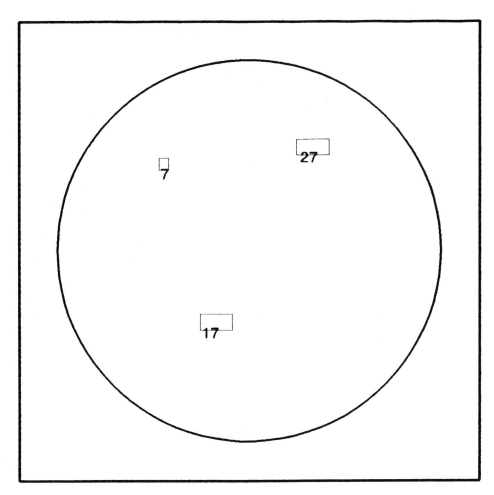

Figure 8.1 The spot-count for each group should be recorded directly onto a simple daily sketch which also indicates the approximate location of each group.

For example:

(1)   11am EST on 1 February is 16:00 UT. (11 hours + 5 hours = 16:00 UT).

(2)   3pm PST on 1 February is 23:00 UT. (3 hours + 8 hours + 12 hours (local time ≥ 1pm) = 23:00 UT).

(3)   3pm in Hawaii on 1 February is 01:00 UT on 2 February. (3 hours + 10 hours + 12 hours (local time ≥ 1pm) = 25 hours − 24 hours = 01:00 UT on the following day).

The Standard Time System is a method of timekeeping whereby all clocks in a specific time zone differ from those in adjacent zones by exactly one hour. The zero zone is that which has the zero degree of longitude running through its center (the *Greenwich*, or *Prime Meridian*). The Earth is divided into twenty-four

Table 8.1

Time zone	Correction
EST	5 hours
CST	6 hours
MST	7 hours
PST	8 hours
Alaska	9 hours
Hawaii	10 hours

similar zones called Standard Time Meridians. Those who are west of the Prime Meridian have their time differences added to the Standard Time of each zone to equal Universal Time, while those to the east of Greenwich are subtracted. Observers who live in areas outside of those included in Table 8.1 can derive the appropriate correction for their location by determining the zone in which they reside, and allowing one hour for each segment.

Image orientation and the determination of the correct hemispherical locations of the solar phenomena are among the most difficult aspects of the observing process for newcomers to solar observation, and these items have been explained previously. Another problem which frequently confronts a new observer is the detection of small, and consequently faint, sunspot groups. These groups are especially difficult to see when they are located near the solar limb (the focus usually requires a slight adjustment if this is the case) or embedded within faculae. Faint spot-clusters often make up a large fraction of the daily sunspot number, and the failure to see them results in the unfortunate combination of a high observer scaling-factor and a low statistical weight.

One clue to the existence of a faint group of spots near the Sun's limb is the presence of the bright-appearing structures known as *faculae*, which often precede a spot's appearance (*facula* is the Latin word for 'torch'). These solar features (Figure 8.2) have been observed for many years, and were first alluded to by Galileo in 1613 in his third letter to Welser (Drake, 1957). They appear as large, irregularly shaped bright areas and are closely affiliated with sunspots.

Faculae seem to be brighter than the surrounding photosphere mostly because they are several hundred kilometers higher, and are less attenuated by the Sun's atmosphere. They are typically fifteen to twenty percent larger in area than their associated sunspot regions. If we exclude the polar faculae that appear in locations which are far-removed from the normal sunspot zones, then faculae are virtually always found near regions of recent sunspot activity, or where spots are about to emerge (Zirin, 1988).

These features are best seen towards the limb where their contrast is higher when compared with the center of the disk, although they can actually be

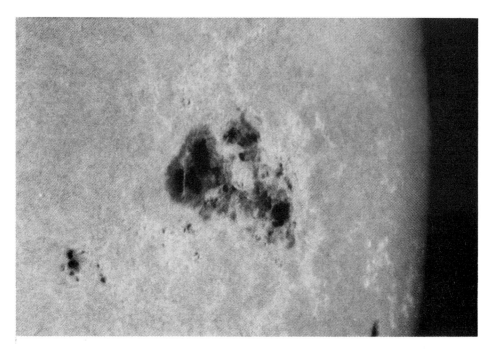

Figure 8.2 The cloud-like features on the Sun called faculae often indicate the impending arrival of a new spot group. Photograph courtesy of Jean Dragesco.

observed all across the visible hemisphere in white-light at wavelengths between 4000 and 4500 ångströms. As we look towards the center of the Sun, our view is directly into a deeper and hotter region of the photosphere, and their detection becomes far more difficult in the normal range of the visual spectrum.

Although faculae have been studied for many years, it is only very recently that high-resolution observations, on the order of 0.3 arc-second, have indicated that portions of these features may be considerably brighter than previously thought. For example, E.V. Kononovich and others at the Shternberg Astronomical Institute in Moscow (Kononovich *et al.*, 1986) have shown evidence that the facular 'filigree' at disk center is fifty to one hundred percent more intense than the neighboring photosphere. This large difference in brightness is hard to explain satisfactorily, but Kononovich has suggested that electrical currents surrounding the magnetic plasma which is associated with faculae and sunspots may be responsible.

Even though both sunspots and faculae are caused by enhanced magnetic fields, it appears that the facular fields are not as strong as those associated with sunspots. As a result they may be less able to inhibit the energy flow from below (Gibson, 1973). Some have suggested (e.g., Schatten *et al.*, 1986) that the energy blocked by the intense magnetic fields of sunspots is in some manner released through faculae.

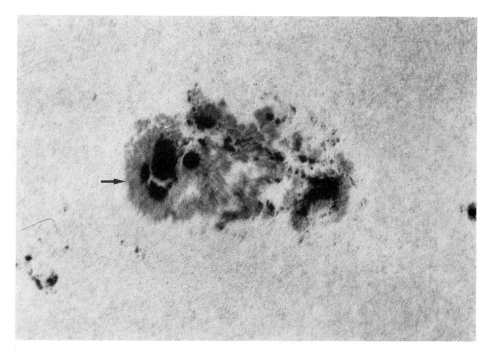

Figure 8.3 New observers often confuse a light-bridge (arrowed) with a solar white-light flare. However, such a flare has a lifetime that is measured in minutes during which time it brightens rapidly and then fades, while observable changes in the light-bridge take place over several hours or days. Photograph by Jean Dragesco.

The *light-bridge* (Figure 8.3) is another common, bright-appearing spot structure. Light-bridges are generally seen as luminous, elongated features which extend into, or divide the umbral areas of large sunspot-complexes. They display a wide diversity of brightness, but often appear to be more radiant than the surrounding photosphere. This is largely an illusion, however, produced by the sudden contrast between the flowing strip of solar 'surface' and the spot's umbra. When observed in white-light in sunspots near the limb, the bridges sometimes appear to be brighter than the neighboring faculae.

Every so often an inexperienced observer will mistake the light-bridge phenomenon for a solar white-light flare. However, it is actually quite easy to tell the difference between these two features: a flare changes its appearance and optical intensity *drastically* in just a very few minutes, while observable changes in the structure of a light-bridge occur over several hours or days. Light-bridges are frequently associated with the onset of the decay of a spot-complex, and occasionally seem to merge with faculae outside of the group. Of course when a bridge clearly divides an umbral area, the spots should be counted accordingly.

When large unipolar sunspots are observed near the solar limb, they appear to be elliptical because of the illusion which results from foreshortening. This misconception is enhanced by another interesting aspect, the depressed appearance of a spot first described in 1769 by the famous Scottish astronomer, A. O. Wilson. Wilson discovered that as a spot approaches the western limb of the Sun, the half of the penumbra that is nearest the edge appears to widen while the other half narrows. The process is a gradual one, becoming more prominent as the spot nears the limb. This phenomenon, which is known as the *Wilson-effect*, is reversed for spots which are rotating onto the disk from the east, and gives the impression that sunspots are sunken areas on the Sun's surface.

The appearance presented by spots near the limb was a mysterious one for the astronomers of the seventeenth and eighteenth centuries. For instance, the effect led both Wilson and the illustrious William Herschel to believe that the Sun's spots were actually rips or holes in the solar atmosphere through which the astronomer could glimpse a dark solar surface (i.e., what we now know to be the spot's umbrae). While not as poetic as the explanations of Herschel and his contemporaries, research (e.g., Bray and Loughhead, 1965) ties the phenomenon to an increased transparency of the spot's umbra, and Zirin (1988) has suggested physical evidence which supports this hypothesis.

Since 1875, and possibly somewhat earlier, grayish 'patches' resembling isolated areas of penumbra have been reported by some observers. They do not appear to fit any of the definitions for other active solar phenomena. These faint irregular areas occur in all areas of the Sun, but are most often seen within ten degrees of the poles or near the center of the disk. They are known as *veiled spots*, a name given to them by the famous French astronomer, Trouvelot, who first observed them in the late nineteenth century (Menzel, 1949). Occasionally they are associated with the remnants of an old decaying group or indicate where a new cluster will soon form. In this way, veiled spots provide the observer with yet another clue to the presence, or impending arrival, of a new spot group.

Wisps are similar to veiled spots in appearance, but they are much more delicate. They are seen as elongated streamers of gray 'haze' which appear as bits and pieces of disconnected penumbra. If observers submit their reports of spot activity to any of the organizations which compile these data, the appearance of either of these phenomena should be noted in the report. Of course, since umbrae are not present in either veiled spots or wisps, they should *not* enter into the sunspot count.

During those times when the Sun is particularly active, a different type of problem often confronts the observer. The difficulty arises when a spot complex suddenly experiences a great surge of growth, when two groups merge, or when one or more new spot clusters erupt very near existing ones. Professional astronomers can analyze these changes by using sophisticated equipment which defines the complex magnetic fields that determine each group's characteristics. Since it is not possible to accomplish this with conventional instrumentation, amateur astronomers must rely on their experience and observations of the past and future development of the region to sort out the situation.

Once an observer becomes familiar with the typical stages of group evolution, he or she can often recognize the changes which occur in the questionable portions of an active area. If in light of these changes it becomes possible to re-divide the region correctly, the previous record(s) should be reconstructed to reflect the actual conditions within the area. This process is much more difficult if the observer has not viewed the Sun for several days or has not maintained appropriate sketches as we have described above.

For these reasons it is very important to view the Sun as often as possible and to keep comprehensive records. In spite of these precautions, a situation will occasionally be encountered that is so complex that even expert observers cannot be certain that they have resolved the problem correctly. In these cases observers must rely on their judgement and experience alone.

It should be realized that since larger telescopes have the ability to resolve fainter spots than smaller instruments do, there is *no absolutely correct* sunspot count for any one day. The concept of the relative sunspot number minimizes this difficulty through the use of a scaling-factor which is based in part upon each observer's instrumentation.

However, there are *incorrect* counts. This situation most often arises when an observer fails to detect some of the fainter groups near the threshold of his or her instrument's resolving ability. To minimize this occurrence, the observer should carefully examine the limb, as well as the higher latitudes and equatorial zones, several times during each observing session. The western limb should not be neglected; new spots can emerge there as well, and the last portion of a group that has just traversed the disk can usually be followed until it is ninety degrees or more from the disk's center.

It is surprising how often a distinct group of sunspots will suddenly become apparent in a location which has just been carefully monitored without their detection. Earlier astronomers often thought that they had witnessed the birth of a spot group when they experienced this effect, but virtually all of those who observe the Sun directly have noted the phenomenon. This impression may relate to some idiosyncrasy of the human eye, or to an atmospheric condition.

Once in a while an observer who views the Sun directly will detect *color* around a sunspot, most often in the penumbral area. The appearance of color near sunspots was quite commonly noted among the great solar observers of the nineteenth century (Webb, 1893). Schwabe, Secchi, Schmidt and Lockyer all reported seeing the effect during the course of their regular observations.

The subject of color in sunspots is an intriguing one, and there are many references to it in the literature (e.g., Webb, 1893; Menzel, 1949; Rosebrugh, 1951). The color is generally perceived as a rosy-hue or perhaps as a violet or brownish-tinge within the group's penumbra. Some reports of color in sunspots may be due to instrumental or physiological factors. Therefore the observer should take care to eliminate any artificial colors which arise from defects in his or her instrument's optical components, particularly within the eyepiece.

89

Still, there are a number of apparently reliable reports of color in sunspots, and this phenomenon probably deserves a more rigorous study than has been undertaken up to now. If color is seen near a spot, the information should be noted on the drawing and included in the observer's monthly report. One interesting explanation for the observation of color within a spot area was given by D.H. Menzel in 1949; he suggested that its appearance might arise from prominence (i.e., filament) activity in the sunspot neighborhood.

Sunspot groups do not move around very much on the Sun; most of their apparent motion is caused by the Sun's rotation. But the Sun, as a gaseous rather than solid body, *rotates differentially*: that is, the various latitudes of the Sun rotate at different rates. The rate is most rapid at the equator and becomes progressively slower towards the poles. As a result the detailed movements of a spot complex are strongly influenced by the rotational rate of a particular latitudinal zone. However, very precise studies have shown that a group's principal spots do increase their separation as the complex evolves to its maximum stage of development.

For example, Waldmeier (1955) found that both the preceding and following spots of a bipolar group show independent motions. (This movement is referred to as the spot's *proper motion*, and is derived by subtracting the effect due to solar differential rotation from measurements of the total movement.) According to Waldmeier, the most westerly spot shows the greatest movement.

On the average, this spot moves *westward* by approximately five degrees of longitude in the six days after the group forms. On the other hand, the following spot moves *eastward* by an average of three degrees during the same interval. The spreading ceases when the cluster reaches its maximum size, and thereafter the proper motion of *both* spots is *easterly*. The rate of this movement is greatest during the groups' initial growth phase.

There is also a smaller movement in latitude. For groups at latitudes greater than ($\pm$) sixteen degrees the motion is poleward, but for lower latitude groups it is usually towards the equator. The measurement of sunspot movement is a difficult problem for an amateur astronomer, though. As has been explained, the differential rotation of the Sun plays an important role in the analysis of a spot's proper motion, and exact rotational rates must be applied to the reduction of equally precise measurements of the spot's movements.

The study and recording of two additional aspects which are related to spot development, Joy's Law (the varying angle of a bipolar group relative to the Sun's equator), and the changing latitude of sunspot emergence during each sunspot cycle known as Spörer's Law, have been discussed in earlier chapters. Each of these effects provides a fertile field of study for the avid observer.

Records of sunspot groups that are large enough to be seen without optical assistance reach far back into history; perhaps to as early as the twelfth century BC. Ancient Chinese observers often mentioned 'flying birds' seen on the solar

Table 8.2

Organization	Address
American sunspot program (AAVSO) (Solar Division)	PO Box 5685, Athens, GA 30604 USA
Astronomical Society of South Africa (Solar Section)	17 Mars Street, Atlasville, Boksburg, 1459 Republic of South Africa
British Astronomical Association (Solar Section)	13 Glencree Park, Jordanstown, Co Antrim, BT37 0QS, Northern Ireland
Oriental Astronomical Association (Solar Section)	1–17 Mikkaichi 1 Chome, Suzuka–shi, Mie–ken 513, Japan
Sonne	Braunfelser Straße 79, W–6330 Wetzlar, Germany
URSA Astronomical Association (Solar Section)	Pilvitie 13G, 37600 Valkeakoski, Finland

disk, and their notes clearly indicate that these markings were large clusters of sunspots (Menzel, 1949).

Many of those who are new to solar observation are especially intrigued by these enormous spot-complexes, and mistakenly regard their appearance as rare events. In reality, groups with areas in excess of one and a half thousand million square kilometers (the visual threshold for a compact group) occur during most phases of the sunspot cycle. Sometimes, as was the case during cycle eighteen, more than five can be seen on the Sun's disk at one time.

These groups, known as *naked-eye spot complexes*, can be safely viewed through a full-aperture solar filter or section of welder's glass of **no less than shade fourteen**, held up to the eye. In order for the data to have any scientific value, the spots should be observed in this manner *before* they are seen telescopically. For, as Sir William Herschel so aptly stated, 'When an object has been discovered by a superior power, an inferior one will suffice to see it afterwards ...' (Todd, 1899). The appearance of naked-eye spots should also be noted on the observer's report.

After the observations for a month have been concluded, a report should be sent to an organization which compiles these data as soon as possible. The names and addresses of several of these associations or their solar sections are provided in Table 8.2. Figure 8.4 is a sample of the report-form which is currently employed by contributors to the American program. The information can be mailed, but if possible it should be sent electronically since it is important that all data are reduced and relayed to the scientific community as soon as possible

SUNSPOT REPORT   MONTH/YEAR

OBSERVER NAME AND COMPLETE ADDRESS

○ REFRACTOR ○ REFLECTOR   APERTURE ( ) In/mm   FL ( ) In/mm

○ OBSERVATIONS DIRECT

○ HERSCHEL WEDGE ○ FILTER   EYEPIECE FL ( ) In/mm

○ OBSERVATIONS PROJECTION

   DIAMETER OF PROJECTED IMAGE ( ) In/mm

○ APERTURE STOP USED DIAMETER ( ) In/mm

DAY	S	T	g	s	R	ng	sg	ns	ss	Remarks
1										
2										
3										
4										
5										
6										
7										
8										
9										
10										
11										
12										
13										
14										
15										
16										
17										
18										
19										
20										
21										
22										
23										
24										
25										
26										
27										
28										
29										
30										
31										

Figure 8.4 A form which is typical of those used for reporting sunspot estimates.

after the end of each month. Information on electronic methods for reporting data to the American program is available from the author at the included address.

Using a Porter Disk to determine heliographic positions

At one time or another, many solar observers will want to derive the heliographic position of one of the Sun's active areas. After this information has been determined, the feature's movement and evolution can be followed more closely, and the activity can be monitored according to its hemispherical location. Many find the constant changes which result from the geometry of the orbital relationship between the Sun and Earth to be an enlightening experience as well. Before we explain how these positions can be calculated, let us briefly review a few of the concepts which will be involved in the reduction process.

Rather than forming a parallel relationship, the terrestrial equator is inclined from the *ecliptic* (the imaginary plane representing the Sun's annual path in the sky) by approximately 23°26′. In a similar manner, the Sun's equator meets the ecliptic plane at an angle of 7°15′. During each year, an observer on the Earth sees a continuous variation in the face of the solar disk which results from the interplay between these angular differences.

In order to determine the orientation of the Sun's disk for a particular day, we should first become familiar with the two series of elements which result from this interaction, P and B_0. These components are shown graphically in Figure 9.1.

The first quantity, P, is the inclination of the North Pole of the solar axis of rotation measured from the north-point in the sky. The second value, B_0, represents the latitude position of the *center* of the solar disk which varies according to changes in the tilt of the solar axis relative to the Earth. Naturally, the latter component appears to decrease to zero at the Sun's limb.

Twice during each year the Earth's equator lies exactly within the plane of the Sun's orbit. At those times B_0 equals zero, and any spot groups on the Sun appear to move across its disk in straight lines. For the remainder of the year the Sun's North Pole is tilted towards or away from the Earth, and the east to west spot-paths take the shape of narrow ellipses.

A third sequence of elements will also be necessary during the reduction process; one consisting of measurements which are equivalent to terrestrial longitude. This series of values is listed in astronomical tables under the heading L_0. It describes the exact longitude of the Sun's central meridian at *zero hours* Universal Time for each day of the year. In a way, when L_0 is 360 degrees it is the

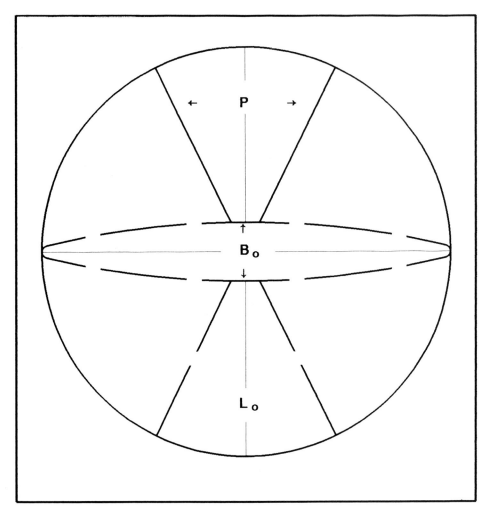

Figure 9.1 The components which are employed in the determination of positions for the Sun's features change considerably during the course of each year. L_0, the value of the Sun's central meridian for zero hours Universal Time, varies continuously as the Sun rotates.

Sun's equivalent of Greenwich, England, where each day on the Earth officially begins. It is important to understand that the amount of L_0 *decreases* each day by nearly 13.2 degrees until the Sun has completed a full rotation. The process then begins again, to be repeated throughout each of the Sun's rotations.

Happily, we do not have to derive individual values for each of these elements. Tables which list them for each day of the year are compiled in the *Astronomical Almanac* and in a number of other astronomical handbooks.

There are many schemes for determining the positions of the Sun's phenomena, but a number of them are so complicated that amateur astronomers understand-

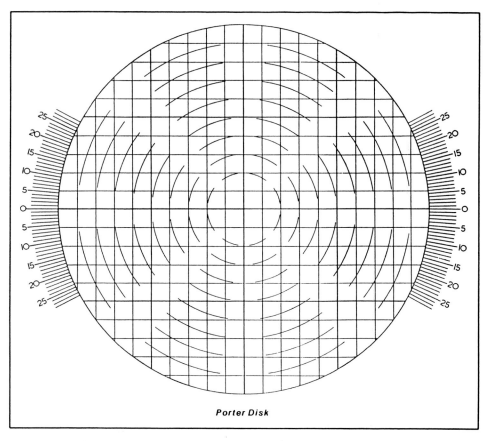

Porter Disk

Figure 9.2 The Porter Disk. The technique for using this disk was devised in 1943 by J. G. Porter as a simple way to derive the heliographic locations of sunspot groups and other solar phenomena.

ably do their best to avoid them. However, there is an easier way, and that is through the use of an ingenious device known as a *Porter Disk* (Figure 9.2). When used with care, Porter's method (Porter, 1943) is accurate to one or two heliographic degrees, about the best accuracy which can be expected for non-professional measurements. At first the method may appear to be complex, but do not be intimidated: in practice it is really quite simple, and employs only simple arithmetic for the few reductions which are required.

In order to apply the techniques, take Figure 9.2 to a good printing company and have it copied onto a sheet of clear plastic acetate to a diameter of fifteen centimeters. You will also need to prepare a drawing of a circle with a diameter which is exactly equal to that of the Porter Disk. The circle should be bisected by two fine lines which divide it precisely into quarterly sections. A supply of drawings can be made up in advance by simply copying the master disk once or twice onto each sheet of paper.

The Sun's image should be projected onto the drawing in a size which is identical in diameter to that of the circle. As has been explained previously, image size is generally controlled by varying the distance between the eyepiece and the projection screen. Since in this situation the observer will be required to maintain an image of constant diameter, the space between these devices will require a slight adjustment as the distance between the Sun and Earth undergoes its annual variation. (As their separation changes, the Sun's angular diameter varies between 31'30".8 and 32'35".2.)

Mount the sketch temporarily on the projection surface, in a position central to the optical axis. With the solar image projected upon the screen and the telescope drive *off*, rotate the drawing about the optical axis so that a sunspot tracks exactly parallel to one of the lines as the spot drifts through the field of view. The direction of travel will always be from the east and towards the west. Thus, the western limb of the Sun will disappear first. Secure the sketch to the screen while maintaining this alignment, and label these points accordingly.

The north-point in the sky for the time of observation can be determined by moving the front of the telescope slightly towards the north, and noting that the southern limb of the Sun will disappear first. Label north and south-points appropriately on the second bisector. The approximate directions which are determined in this way will not change substantially unless the observer uses a different type of instrumentation, alters the method of observation or changes his or her location. As a result, they need to be determined only once; at the time of the initial observation.

Start the telescope's drive mechanism and center the image. Maintain a sharp focus as you sketch the principal spots in each group onto the drawing, tracing their outlines with a sharp pencil. The fainter clusters will become more apparent if one gently taps the projection screen, or the instrument itself. The date and Universal Time of the observation should also be noted on the drawing.

After the observations have been completed remove the sketch and superimpose the Porter Disk transparency so that the two diameters coincide, Look up the value of *P* for the date of observation from the appropriate table. With the aid of the degree-scale at the sides of the disk, rotate the transparency about its center to reflect this angle. Rotation should be from *north towards west* if *P* is negative, and from *north towards east* when the angle is positive.

The Porter Disk is divided by solid lines into twenty parts along each axis. These divisions should be used to measure values for *x* (along the east–west axis) and *y* (along the north–south axis) for each feature, to an accuracy of two decimal places. When measuring the position of a bipolar sunspot group, use the center of the principal spot (often the leader, or most westerly spot) as a reference point; it will generally outlast other spots within the complex. Otherwise measure to the center of a unipolar spot or other feature.

Apply the following law of signs to these quantities:

> Component *x* is taken as positive west of the center-line, and as negative to the east.

Table 9.1

d	$B_0{}^\circ$	0	1	2	3	4	5	6	7
0.0		0.00	0.02	0.03	0.05	0.07	0.09	0.10	0.12
0.1		0.00	0.02	0.03	0.05	0.07	0.09	0.10	0.12
0.2		0.00	0.02	0.03	0.05	0.07	0.09	0.10	0.12
0.3		0.00	0.02	0.03	0.05	0.07	0.08	0.10	0.12
0.4		0.00	0.02	0.03	0.05	0.06	0.08	0.10	0.11
0.5		0.00	0.02	0.03	0.05	0.06	0.08	0.09	0.11
0.6		0.00	0.01	0.03	0.04	0.06	0.07	0.08	0.10
0.7		0.00	0.01	0.02	0.04	0.05	0.06	0.07	0.09
0.8		0.00	0.01	0.02	0.03	0.04	0.05	0.06	0.07
0.9		0.00	0.01	0.02	0.02	0.03	0.04	0.05	0.05
1.0		0.00	0.00	0.00	0.00	0.00	0.00	0.00	0.00

Component y is taken as positive north of the center-line, and as negative to the south.

The transparency is also divided into a series of broken, concentric circles which are plotted at tenths of the disk radius, beginning at 0.2 radii. These markers are employed to provide a simple estimate for d, the distance of the spot or feature from the center of the disk.

The heliographic latitude, b, will then be given by,

$$\sin b = y + \text{correction value.}$$

The appropriate correction value for a given d and B_0 pairing is provided in Table 9.1, and is awarded the same (positive or negative) sign as the tabular value of B_0 on the day of the observation. Particular attention should be paid to the signs of y and the correction value when solving the equation. The latitude can then be read directly from Table 9.2.

The computation of the heliographic longitude begins by deriving l according to

$$\sin l = x(\sec b),$$

with $\sec b$ again read from Table 9.2. Following this solution, l is also read from Table 9.2.

The longitude of the Sun's central meridian at zero hours Universal Time on the day of the observation (L_0) is taken from the *Astronomical Almanac*. This value should be refined through the process of linear interpolation so that it accurately reflects the Universal Time of the observation, *before* it is included in the following equation. When interpolating, remember that L_0 decreases by approximately 13.2 degrees per day.

Table 9.2

Angle(°)	sin	sec	Angle(°)	sin	sec	Angle(°)	sin	sec
0	0.00	1.00	20	0.34	1.06	40	0.64	1.31
1	0.02	1.00	21	0.36	1.07	41	0.66	1.33
2	0.03	1.00	22	0.37	1.08	42	0.67	1.35
3	0.05	1.00	23	0.39	1.09	43	0.68	1.37
4	0.07	1.00	24	0.41	1.09	44	0.69	1.39
5	0.09	1.00	25	0.42	1.10	45	0.71	1.41
6	0.10	1.01	26	0.44	1.11	46	0.72	1.44
7	0.12	1.01	27	0.45	1.12	47	0.73	1.47
8	0.14	1.01	28	0.47	1.13	48	0.74	1.49
9	0.16	1.01	29	0.48	1.14	49	0.75	1.52
10	0.17	1.02	30	0.50	1.15	50	0.77	1.56
11	0.19	1.02	31	0.52	1.17	51	0.78	1.59
12	0.21	1.02	32	0.53	1.18	52	0.79	1.62
13	0.22	1.03	33	0.54	1.19	53	0.80	1.66
14	0.24	1.03	34	0.56	1.21	54	0.81	1.70
15	0.26	1.04	35	0.57	1.22	55	0.82	1.74
16	0.28	1.04	36	0.59	1.24	56	0.83	1.79
17	0.29	1.05	37	0.60	1.25	57	0.84	1.84
18	0.31	1.05	38	0.62	1.27	58	0.85	1.89
19	0.33	1.06	39	0.63	1.29	59	0.86	1.94
						60	0.87	2.00

Note: Values for angles greater than $60°$ are not included in the table. The advanced foreshortening near the Sun's limb produces a significant decrease in the accuracy of position for features which are located outside of this zone.

The correct heliographic longitude is then given by,

$$\text{heliographic longitude} = l + L_0.$$

An estimate of the area of a sunspot-cluster can also be made with the Porter Disk. Each square unit on the device contains sixty-four square-millimeters. On the recommended scale, one square millimeter near its center represents an area of approximately twenty-eight (28.3) millionths of the visible solar hemisphere.

To compute a group's area, lay the transparency over the tracing and carefully estimate the number of units within the cluster's border (be certain to include the penumbra). Count all units which are covered by at least half, and ignore the others. Multiplying this result by the number of millionths contained within each square millimeter gives the approximate area for groups located close to the center of the disk. However, foreshortening will distort the measurement of groups away from the central zone, and consequently this result should be adjusted by multiplying the quantity by the appropriate secant from Table 9.2. The correct value for the latter function will be found in the table opposite the sine which is equal to measurement d (see example).

One square degree on the surface of a sphere contains approximately forty-eight and a half millionths of its hemisphere. Thus, each square millimeter on the Porter Disk represents a heliographic area which is slightly greater than a half of one square degree. For example, for a projection made 24 January 1988 at 13:00 hours Universal Time. Say,

$$P = -8°.5; \quad B_0 = -5°.3; \quad L_0 = 315°.5.$$

The Porter Disk is placed over the completed drawing and the transparency is rotated from the north point towards the west through $8°.5$. (Rotation is towards the west since 'P' is negative.)

The coordinates of a spot are found to be (for example):

$$x = 0.26; \quad y = -0.37; \quad d = \sim 0.45.$$

The correction from Table 9.1 is $(-)0.09$. (The correction is assigned a negative value since 'b_0' for the date of the observation is negative.)

Then:

$$\sin b = -0.37 - 0.09 = 0.46(-) \quad \text{and} \ b = -27°.4 \text{ (south)}.$$
$$\sin l = 0.26 \times 1.13 = 0.29 \quad \quad \text{so} \ l = 16°.9$$
$$L_0 = 308°.4 \ (\underline{\text{after}} \text{ interpolation}),$$
$$\text{longitude} = 325°.3.$$

Because d in our example is equal to 0.45, the group's area should be multiplied by the secant opposite sine 0.45 (1.12) to adjust for foreshortening. Values for all necessary functions and angles are included in Table 9.2.

10

Observing solar white-light flares

As fortune would have it, the first solar flare to be observed at any wavelength was also one of the rarest of all astronomical events, a *solar 'white-light' flare*. The phenomenon was observed visually in 1859 by the English astronomer and brew-master, Richard Carrington (Carrington, 1859). In an extraordinary stroke of luck, the same event was also seen independently by a second English observer, Richard Hodgson. The flare was so intensely bright that at first Carrington thought that the screen attached to his instrument's lens had shattered; but the brilliant patch of light was clearly in physical association with the Sun (Todd, 1899).

The flare lasted for only a short while, and Carrington is said to have been quite perplexed when he returned to his telescope with a friend to witness the event, only to find that it had all but disappeared. The flare was followed some sixteen hours later by the strongest geophysical disturbance in over a century, and by dazzling aurorae. As one might imagine, the inferred association between the flare and the effects seen at the Earth's surface stimulated a considerable interest in the solar–terrestrial relationship.

Occasionally a very energetic flare penetrates the solar chromosphere and extends from the Sun's atmosphere down into the photospheric region. As a consequence, the flare can be seen in the visual region of the spectrum (between 4000 ångström and 7000 ångström), and the phenomenon is referred to as a white-light flare (WLF). The sighting of such an event is a very unusual occurrence, for less than eighty have been recorded since the Carrington–Hodgson discovery.

Even though WLFs occupy only a small percent of the area encompassed by hydrogen-alpha (Hα) flares such as the one shown in Figure 10.1, they are tremendously energetic events which often carry with them far-reaching geophysical implications. They may radiate nearly as much energy in only a few moments as the total emitted by their Hα counterparts (i.e., $\sim 10^{34}$ erg) with peak power near 10^{28} erg per second. One of the most intense WLFs ever recorded is shown in Figure 10.2.

For a number of years, the National Solar Observatory at Sacramento Peak has coordinated an effort to record these events and to compile information about them. The WLF observing program at Sacramento Peak Observatory relies on a

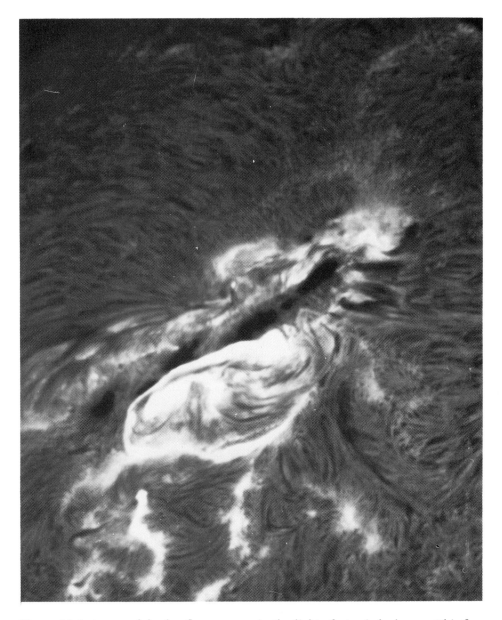

Figure 10.1 A powerful solar flare as seen in the light of atomic hydrogen. This fine photograph was provided by Thomas G. Compton.

small-aperture patrol instrument known as the 'Multi-band Polarimeter.' The instrument has been in regular use since 1980 and is scheduled for upgrade in the near future. Those interested in the details of its configuration can obtain a fairly comprehensive description in Neidig (1983).

Dr Donald F. Neidig directs the white-light flare program at Sacramento Peak

WHITE-LIGHT FLARE
24 APR 1984 (1901 EST)
OPTICAL/X-RAY CLASS 3B/X13

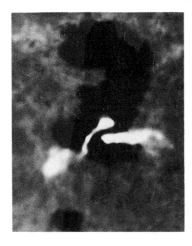

3610 Å

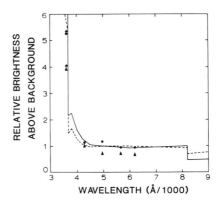

IONOSPHERIC DISTURBANCE CAUSED
BY THIS FLARE NEGATED CRITICAL
HF COMMUNICATION LINKS FOR 2 HRS

Figure 10.2 One of the strongest solar white-light flares ever recorded occurred on 24 April, 1984. This photograph of the event was taken at the National Solar Observatory at Sacramento Peak, and was supplied through the courtesy of D. F. Neidig.

Observatory. Much of the following statistical information was generously supplied by Dr Neidig (Neidig and Cliver, 1983) or taken from Neidig (1983, 1988).

Seventy percent of the WLFs which have been recorded since the Carrington–Hodgson event have occurred in the Northern Solar Hemisphere. They have appeared at an average latitude of eighteen degrees in the Northern Hemisphere and at about thirteen degrees in southern locations. According to these data, WLF activity in the north may be expected to begin one or two years *before* the sunspot cycle reaches its maximum strength.

On the other hand, WLF activity in the Southern Hemisphere usually begins approximately one year *after* maximum. When the hemispheres are combined overall activity appears to peak at about the same time that WLF activity commences in southern locations. The relationship between WLFs and the occurrence of sunspot cycle maxima is shown in Figure 10.3.

Because of their relatively low contrast when viewed against the solar background, WLFs are detected only rarely during each sunspot cycle. The probability of such a sighting, while admittedly low, is highest during and just after the maxima of strong sunspot cycles. The flares often appear as one or more small, rapidly brightening *patches* or *ribbons* near or within a large and complex sunspot group. The events are visible only during a very brief interval (usually

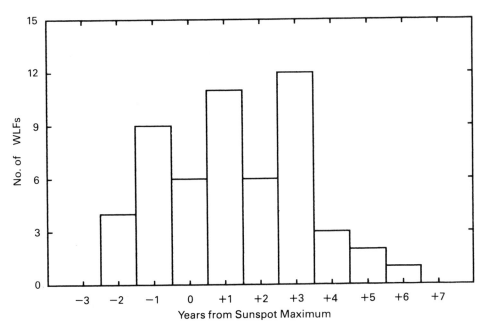

Figure 10.3 Research has shown that the great energy bursts known as white-light flares (WLFs) usually occur a year or more before or after sunspot cycle maximum, rather than at its peak.

between one and ten minutes) but exhibit a brightness which can exceed 300 percent of that which is attributed to the surrounding photosphere.

The sunspot complexes which are associated with WLFs almost always encompass areas of more than 500 millionths of a solar hemisphere (around one and a half thousand million square kilometers) and there is a high probability that the group will be a large, visually bipolar complex with a longitudinal spread in excess of ten degrees. Those with many small sunspots between the principal leading and trailing spots are especially likely sources. Groups with these characteristics have been described previously.

Above all, the most likely area for a WLF is adjacent to, or within, the 'K' type penumbra of the principal spot of a 'delta' magnetic-class sunspot group (Table 10.1). This particular class of sunspot penumbra has a diameter greater than two and a half heliographic degrees, is asymmetric or oval, and encloses two or more umbrae. A typical group with penumbra of this class is shown in Figure 10.4. When the penumbra exceeds five degrees in longitudinal spread it almost always includes both magnetic polarities, and the group is given the magnetic classification gamma. A second designation, delta, is appended to the group's magnetic classification when opposite magnetic polarities exist within two degrees of one another inside the single penumbra. These delta groups especially, and to a lesser extent the gamma types, are statistically the most likely sources of the

Table 10.1   *Characteristics of sunspot groups associated with WLFs (August 1972–December 1982)*

WLF date	Magnetic class	Group class	Area (msh)[a]
2 August 1972	delta	EKC	1050
2 August 1972	delta	EKC	1050
7 August 1972	delta	EKC	910
4 July 1974	delta	FKC	1020
10 September 1974	delta	DKC	750
9 July 1978	delta	DKO	760
10 July 1978	delta	EKC	1080
11 July 1978	delta	EKC	1330
3 June 1980	delta	DKO	310
4 June 1980	delta	EKI	990
1 July 1980	beta	EKI	400
27 January 1981	beta–gamma	DKI	520
26 February 1981	gamma–delta	CKI	520
24 April 1981	delta	DKC	1170
13 May 1981	delta	FKC	1510
26 July 1981	beta	FKI	2140
4 June 1982 (13:30 UT)	delta	EKC	930
4 June 1982 (14:23 UT)	delta	EKC	930
5 June 1982	delta	FKC	1160
6 June 1982	delta	EKI	1180
25 June 1982	delta	EKI	940
26 June 1982	beta	EKI	560
15 December 1982	beta–gamma–delta	DKI	270
17 December 1982	beta–gamma–delta	DKI	500

[a] Millionths solar hemisphere.
Source: From information supplied in Neidig and Cliver, 1983.

strongest solar flares; events which occur as the spot's magnetic qualities clash and attempt to rearrange themselves into more traditional patterns.

Dr Neidig provides the following information for potential white-light flare observers (Neidig, 1988):

> The following data for a WLF are *essential*: The date, Universal Time of maximum brightness, active region identification, and information on how the observation was made (especially the wavelength, in cases where filters are used).[1] In addition, drawings or other description of the

[1] Rather than projection techniques.

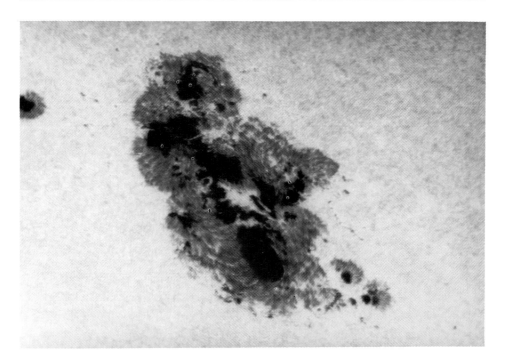

Figure 10.4 This huge magnetically complex (delta class) sunspot group was responsible for the many strong flares which took place during March 1989. Note the compact asymmetrical appearance of the spot group; a description which is typical of groups which produce intense solar flares. Photograph courtesy of D. F. Neidig.

flare in relation to the sunspot environment are of interest. Another item, although difficult to obtain, is an estimate of the flare's peak brightness compared with the photosphere outside (but near) the active region where the flare occurs.

Please note that recent studies suggest that *all* flares are WLFs. The reason that so few have been reported is simply a matter of signal-to-noise ratio, i.e., the difficulty of detecting (usually) weak continuum against the bright photospheric background. Quantitative data, of course, are preferred, but many WLFs are missed by major observatories, and in these cases it is very nice to have qualitative information from skilled observers.

Although they are quite rare, the exceptional brightness exhibited by a white-light flare allows it to be seen quite easily with a small telescope. A blue filter with a peak transmission near 4300 Å and a passband width less than 100 will increase the visual contrast of the flare compared with the photosphere, and thus aid the detection process. A Wratten 47 blue filter is one simple and inexpensive choice for this purpose (Compton, 1987). Since the blue filter offers **no**

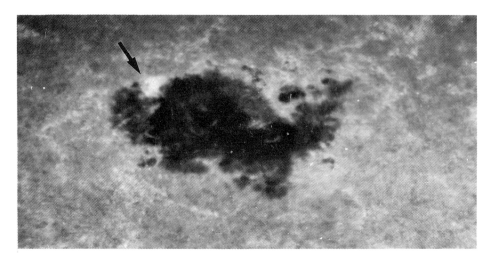

Figure 10.5 The appearance of a white-light flare (arrowed) is somewhat subdued when it is viewed in the normal visual portion of the spectrum. This photograph was taken at Sacramento Peak Observatory, within a band centered on 4275 ångströms.

protection from solar radiation by itself, it **must be used in conjunction with the observer's normal solar filtering.**

A second, although slightly more expensive solution is to obtain a narrow passband glass filter in this ångström range; one with a band-width of a few tens of ångströms. Filters of this description can frequently be purchased for a fraction of their original cost through the larger scientific supply companies which deal in surplus or end-lots of these optical items. Figure 10.5 should provide the observer with a general idea of the appearance of a very bright WLF when it is viewed in the shorter-wavelength portions of the visual spectrum.

Observers who monitor solar activity photographically or through video-tape techniques, should use filters that peak below 4000 Å, generally in the neighborhood of 3600–3800 Å, which will increase a flare's contrast considerably. Again, even narrower passband-width filters (widths on the order of a few ångströms) which contain large numbers of absorption spectra are the most satisfactory. According to Dr Neidig, 'A bluish color has been noted in several visual observations of WLFs. Apparently this is due to the rapid drop in intensity of the solar background at short wavelengths, combined with an actual increase of flare emission in the blue.' However, visual observations should not be attempted shortward of 4000Å.

A number of amateur observing groups compile information about white-light flares, and those who regularly view the Sun should always be alert for the occurrence of such flares. The American program is cooperating with Dr Neidig and Sacramento Peak Observatory in the compilation of these unusual observations. Consequently, observers are strongly encouraged to report their sightings of a WLF directly to the author at the address included in the preliminary pages of this book. A detailed report which includes the information requested by Dr Neidig is required. After evaluation, a verified observation may be included in the Sacramento Peak Observatory catalogue of these events.

11

Detecting solar flares electronically

In the previous chapter we discussed a very rare type of solar flare which is so bright that it can be seen without any specialized filtering other than that required for viewing the Sun safely. However, large numbers of other flares occur each day when the Sun is active, and while most are not as spectacular as a white-light flare, they too are among the most interesting and exciting events in nature.

Flares occur when strong magnetic fields, extending high into the Sun's atmosphere above sunspots or other portions of the photosphere, suddenly collapse and then recombine into simpler structures. In the process, vast amounts of energy are frequently released into the interplanetary medium. Although flares have been studied for well over a hundred years, it remains a mystery how the collapse of a magnetic field can release more energy in just a few moments than the United States requires for an entire year. Viewed differently, Noyes has stated that the largest solar flares have total energy emissions which can equal that released by *two and a half thousand million* hydrogen bombs! (Noyes, 1982).

The development of a powerful flare and its associated mass ejection is shown in Figure 11.1, a spectacular series of photographs taken in the red spectral line of atomic hydrogen at National Solar Observatory, Sacramento Peak. The large, complex sunspot group which produced the flare during March 1989 is pictured in the previous chapter (Figure 10.4).

The majority of our information about solar flares has been gained from optical observations made with hydrogen-alpha filters (Figure 11.2), and through studies of flare spectra. Recently, however, new information has been gained from satellites such as the Geostationary Operational Environmental Satellite (GOES), which are equipped with special detectors that monitor the far-ultraviolet and X-ray radiation from flares in the one to eight ångström range.

The incidence of solar flares and studies of their intensities are of increasing interest as we send astronauts farther into space for longer and longer periods of time. The radiation which is associated with these enormous bursts of energy can pose serious problems for humans as well as for their instrumentation, and it is especially frequent and intense for a few years before and after the maximum of each sunspot cycle. Since the geomagnetic disturbances which are a byproduct of strong flare activity also affect radio communication, a number of

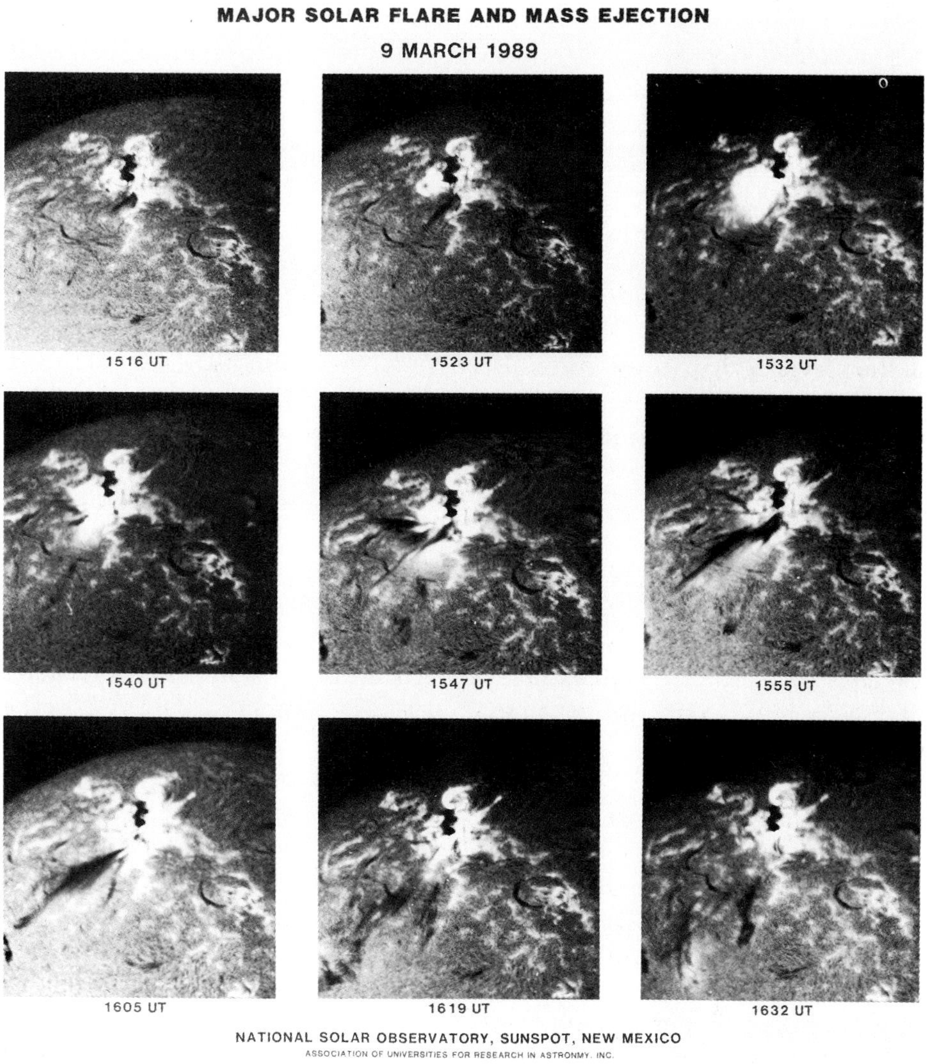
MAJOR SOLAR FLARE AND MASS EJECTION
9 MARCH 1989
1516 UT
1523 UT
1532 UT
1540 UT
1547 UT
1555 UT
1605 UT
1619 UT
1632 UT
NATIONAL SOLAR OBSERVATORY, SUNSPOT, NEW MEXICO
ASSOCIATION OF UNIVERSITIES FOR RESEARCH IN ASTRONOMY, INC.

Figure 11.1 This spectacular series of photographs taken at Sacramento Peak Observatory details the eruption and subsequent mass-ejection of an intense flare. The huge, spray-like ejection of dark material shown at 15:55 UT is more than 200 000 kilometers in length. The huge bursts of energy which accompany these events can be detected indirectly through their effects upon the ionosphere, and occur simultaneously with the flare's visual appearance. D. F. Neidig generously supplied this photographic series.

military and civilian agencies are concerned with the effects of flares on the Earth's environment.

Some interesting information (Zirin, 1988) concerning the energy and optical classifications of solar flares is provided in Table 11.1

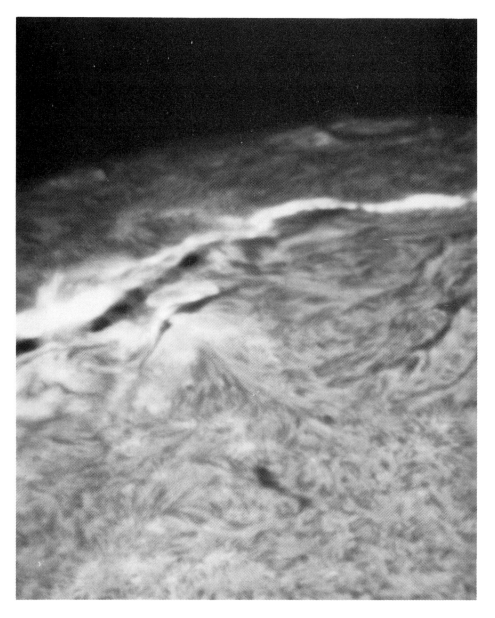

Figure 11.2 An intense flare seems to float above the photosphere in this superb photograph supplied by Thomas G. Compton.

An *f* (faint), *n* (normal) or *b* (bright) is generally appended to the optical brightness class value.

Flares with estimated energy bursts as high as X20 have been recorded. However, only a comparatively small number of the brightest flares occur during each of the Sun's activity cycles. For instance, Smith and Smith (1963) have

110

Table 11.1 *Flare parameters*

Area (10^{-6} solar hemisphere)	Optical class	Typical flux	NOAA class
⩽ 200	S	5	C2
200–500	1	30	M3
500–1200	2	300	X1
1200–2400	3	3000	X5
>2400	4	30 000	X9

shown that the number of optical class 3 or 4 flares at sunspot maximum averages about twenty, in contrast to the thousands of lesser-intensity events which are normally recorded near a cycle's peak.

I am often asked where on the Sun flares are most likely to originate, a question which does not have a simple answer! Above all though, flares occur in regions which have already produced them. That is, if a group has previously yielded a number of flares it may well continue to do so. Growing spot-clusters which are magnetically classed as gamma or delta types are also very likely to produce flares, especially spot-clusters which contain several dark umbra within a single large penumbra.

A considerable portion of a flare's energy is released in the ultraviolet and X-ray regions of the spectrum, and does not penetrate to the surface of the Earth. Instead, much of this radiation is absorbed in the atmosphere, high above the Earth's surface. Certain flares also emit clouds of highly energetic cosmic rays (mostly protons) which travel to the Earth in as little as fifteen minutes. Along with other atmospheric and ground-level disturbances, these often produce disruptions centered at the poles known as *Polar Cap Absorptions*.

A good deal of this energy is dissipated in the ionosphere, a section of the upper atmosphere which is strongly influenced by solar radiation. The lowest part of the ionosphere, a region which is known as the *D-layer* (see Figure 5.1), forms above the daylight hemisphere each day at sunrise at an altitude of about seventy kilometers. Since the D-layer results from the normal effects of the Sun's radiation on its components, it is very sensitive to changes in the amounts of radiation. Ionization increases dramatically when a flare erupts, but the free electrons which result from this process are recombined almost immediately when the flare ends. Still, the effect of this sudden increase in radiation on the atmosphere is very pronounced while the flare is in progress.

The *E-layer* is much less sensitive to flare effects than are the other atmospheric layers. The highest region of the ionosphere, the *F-layer*, is affected similarly to the D-layer; but at its altitude (the F-layer begins at an average height of approximately 300 kilometers) the atmospheric density is so low that the entire process

111

is slowed, and flare effects may not appear for a day or more (Chernan, 1978). However, all four layers of the ionosphere (the F-region actually separates into two layers during the daytime) are influenced to some extent by flare emission.

The D-layer is actually the means of propogation for low and *very low frequency* (VLF) radio signals. Radio signals in this frequency range are propagated in a ducted mode in the space between the D-layer and the Earth's surface, after the manner of a ground wave. This 'wave-guide' becomes a more efficient propagation medium when the conductivity of its upper surface is enhanced by the extra ionization produced by a flare event. In practice it is this enhancement, which occurs at almost the same moment as the flare's optical appearance, that can be easily detected electronically.

Since the waves are propagated in this manner, it has long been realized that low frequencies offer a way to maintain radio communications at high latitudes when the magnetic field experiences severe storm conditions and the higher frequencies are disrupted. In addition, VLF radio waves have another very interesting property. That is, they can scatter into a conducting medium such as salt water for a fraction of a wavelength.

For this reason VLF signals are used the world over by nations that have submarines, which trail antennas just beneath the ocean's surface enabling them to receive messages while remaining submerged. This system of communication has resulted in a world-wide collection of extremely powerful radio transmitters which usually operate continually. Consequently, the network provides a convenient VLF signal source which can be used to monitor solar flares from anywhere in the world.

The first organized attempt to record flare activity in this manner resulted from the distribution of a circular prepared in 1955 by M. A. Ellison of the Royal Observatory in Edinburgh. The circular described the influences of strong flares on the atmosphere, and the construction of an elaborate vacuum tube receiver to chronicle the activity. The receiver would monitor these phenomena by recording an atmospheric effect originally unearthed by the French physicist R. Bureau, and known as a *sudden enhancement of atmospherics*, or SEA.

The SEA method successfully employs the constant stream of static generated by lightning strikes around the world as a signal source. The static from each strike produces a 'pulse' which can be detected electronically near a frequency of twenty-seven kilohertz. Strong flares on the Sun can emit large quantities of X-rays which travel to the Earth at nearly the speed of light, increasing the ionization of the D-layer and causing this signal to be strengthened considerably. In the SEA method, these enhancements are usually recorded on small strip-chart recording devices, and are then measured according to their duration and intensity.

After the appearance of Ellison's article, the receiver was re-designed into a small transistorized circuit for use by a group of observers working within the American Association of Variable Star Observers' Solar Division. As the development of the technique proceeded and its effectiveness became clear, the Chairman of the US International Geophysical Year (IGY) Solar Activity Panel, Dr Walter Orr Roberts, urged the establishment of a network of receiver–recorder systems as a part of the IGY program of flare investigation. Four recording devices were provided to the group by the National Bureau of Standards, and shortly thereafter the program began to contribute its data to the world-wide scientific community; an effort which continues today.

However, rather than recording the atmospherics produced by lightning, the majority of today's observers monitor VLF radio stations for a similar effect called a *sudden enhancement of signal*, or SES. SES effects are observed on field-strength recordings of very stable VLF transmissions. They are similar to SEA, but since the receivers are tuned to man-made sources the interference and variation in signal associated with the older SEA method no longer poses a significant problem. As a result the receiving stations can be located in urban as well as rural areas.

The effort to detect these events continues to be highly encouraged by the National Oceanic and Atmospheric Administration (NOAA) since the information gained is useful for the purposes outlined above, and also because it plays a vital role in the preparation of radio propagation and other forecasts. In light of these needs, the American program supplies NOAA with a monthly analysis of the data obtained by members of its international flare-patrol network. These results appear regularly in *Solar-Geophysical Data*, and in a number of other publications which are devoted to the study of the solar-terrestrial relationship.

A large percentage of the Sun's flares can be detected by the SES/VLF method, and the technique is well within the means of an amateur astronomer. Participants use simple transistorized radio receivers which usually operate at frequencies between five and fifty kilohertz. Like the SEA method, the receiver generally records changes in the integrated signal strength onto special pressure-sensitive paper-tape which is stored within a small recording device.

During the night the recorder typically draws an erratic line towards the top of the recorder's scale. However, as the Sun rises a precipitous drop in intensity occurs, followed by a fairly steep rise and 'hump.' One explanation for this strange pattern was offered in 1960 by International Telephone and Telegram Corp. technical supervisor, David Warshaw. According to his scenario (Warshaw, 1960) the dip may occur when an 'ionized band' is positioned directly between the receiving station and the reflective E- and F-layers, and absorbs the signal for a few minutes. Since the D-layer is lightly ionized during the early morning hours, the hump which follows could result from a sudden improvement in signal enhancement. This phenomenon is known as the 'sunrise effect,' and is shown in Figure 11.3.

If the Sun is not active, the recorder draws a slightly bowed line throughout the day as the ionization level of the D-layer rises to a maximum around local noon, and then gradually decays. When a flare does occur, however, the signal

113

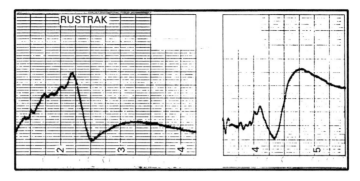

Figure 11.3 The 'sunrise effect.'

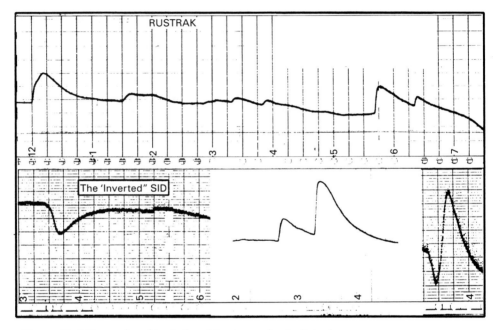

Figure 11.4 Examples of typical sudden ionospheric disturbances.

strength is affected in the described manner, and this results in a sudden rise on the otherwise straight trace.

The enhancement generally has a characteristic shape which makes it quite easy to recognize: a sharp rise to maximum level, followed by a rather slow decline. Typically, the ascent to maximum occurs in just a few minutes, and the enhancement decays completely in about one hour. Several events of differing strengths are shown in Figure 11.4. The atmospheric enhancements which result from these effects are generally referred to as *sudden ionospheric disturbances*, or simply, SIDs. They were originally described by Mögel (1930) and by Dellinger (1935).

114

Table 11.2 *SID importance*

Duration of event	Importance rating
18 or less minutes	1−
19 − 25 minutes	1
26 −32 minutes	1+
33 − 45 minutes	2
46 − 85 minutes	2+
86 − 125 minutes	3
126 or more minutes	3+

When flares are detected through this technique, the resulting SIDs are rated according to two parameters (Lincoln, 1964; Ammons, 1981; Coffey, personal communication): their *importance*, and their *definiteness* on the recording. The importance is rated on a scale of 1− to 3+, with 3+ representing the most intense event. The rating numbers assigned to SIDs are based upon a flare's duration, and result from an averaging of the subjective importance for all stations that record the event. In general, importance is selected according to the graduated scale provided in Table 11.2.

When a SID is analyzed, it is customary to calculate its duration by first determining the exact Universal Time that the event begins and ends. The time of onset is taken as that instant when the recorded trace just begins to rise. The end-time is more difficult to determine, but it is usually recorded as the time that the declining branch of the SID intersects the diurnal trend-line. The descending branch may decay smoothly into the trend-line, or on occasion it may 'drop' suddenly, ending the event.

The time of event-maximum is also measured. It is normally assumed to be the moment when the rising branch of the SID first begins to level off, rather than the time of greatest amplitude. At times these two conditions occur simultaneously and the event has a sharply defined peak, although this is generally not the case. When any of the required times are impossible to determine, a 'U' is appended to the last distinctly recorded time. On the other hand, when events occur in rapid succession (a second event begins before the previous SID ends) a 'D' is added to the end-time of the first event and the start-time for the second SID is presumed to be equal to it.

Although the typical SID appears as we have described (the signal increases abruptly and then gradually returns to normal), others may appear to be inverted or differ in other ways. The inverted SID has the same general appearance as the usual event with the single exception that it is inverted. That is, the signal's strength decreases suddenly and then gradually returns to normal.

Occasionally a third type of signal enhancement will be encounted. In this case, the signal first decreases very sharply, and then rapidly increases before

Table 11.3 *SID definiteness*

Event criteria	Definiteness rating
Questionable	0
Possible	1
Fair	2
Reasonable	3
Reasonably definite	4
Definite	5

slowly returning to normal. Mitra (1970) described similar event-shapes in an analysis of changes in the long-wave field intensity which are experienced during a flare, and attributed the differences to the wavelength and intensity of the X-rays which are emitted by the flare.

The definiteness is a confidence-rating on a scale of 1 to 5, with 5 the most distinct. Definiteness carries a two-fold meaning. The first refers to the clarity of an event on an individual recording, and the second deals with the number of stations that record a SID. Generally, the more observers that record the event, the higher the rating number that is assigned. A small indistinct flare can be detected by several observers, thereby giving the event a higher rating. A general guide to definiteness is provided in Table 11.3.

It is not difficult to equal, and occasionally surpass the sensitivity of the X-ray detectors on satellites, and SID observers often find clearly recorded events that satellite monitors appear to have missed. The D-Layer is very sensitive to the influences of flare radiation, and even small flares can enhance its free electron content sufficiently to produce a recordable event. When the Sun is very active, a dozen or more SIDs can be recorded in a single day.

There is always a need for additional flare-patrol members, and the program would be very pleased to hear from anyone who is interested in participating in this type of activity. The monitoring system is inexpensive and easy to fabricate. (The entire recording station, consisting of receiver, antenna and recorder, can be assembled for less than the cost of a moderately sized reflecting telescope.) The equipment operates automatically and is not seriously impacted by weather conditions, so it is possible to monitor the Sun for flares continuously during the day even though the 'observer' may not be present.

Recently, an additional reason for the monitoring of VLF transmissions according to these techniques has arisen, concerning the detection of what are thought to be *gamma-ray bursts*. It has been reported (Fishman, 1988) that a similar burst in 1983 may have instigated a change in the amplitude of a VLF signal from Rugby, England which was recorded in Antarctica. According to Dr Fishman, events such as this may be caused by explosions on, or near, neutron stars.

On the recording, such events have shapes which while of less amplitude and duration, are similar in appearance to normal SID events. Since the ionization level decreases significantly after the Sun has set, night-time transmission paths may offer the best opportunity to register these effects. Only a few will be strong enough to be recorded each year, but their importance offers a second, and potentially valuable use for a monitoring station.

In the following chapter, we will describe the construction of a reliable receiver and discuss the simple equipment and procedures which will allow it to be put to use gathering important scientific data. It is helpful if those who wish to participate in this project also have some experience with electronics, but these are very simple devices which do not require a great deal of expertise to build or operate. Advice is always available through the program's Technical Coordinator, and from other members of the patrol.

12

Constructing a solar flare monitoring station

Radio receivers which are capable of monitoring stations that operate at between fifteen and one hundred kilohertz are generally used to detect the sudden enhancements of signal (SES) which are generated by the atmospheric effects of solar flares. Research into the propagation of very low frequency (VLF) radio signals has indicated that the ionosphere is most responsive to a frequency of between twenty-four and twenty-seven kilohertz (Chernan, 1978). Consequently, the receiver described here has been designed to operate in the range of twenty to thirty-two kilohertz: a band of frequencies which includes a number of VLF stations that are particularly well-suited to the detection of solar flares.

The receiver can actually be thought of as a 'dual purpose' unit, in that it can be adjusted to monitor either sudden enhancement of atmospherics (SEA) or SES effects. However, the SES technique should be chosen in this instance, and the unit tuned to a man-made signal source (a VLF radio station). In general, the quality of the data acquired in this manner is superior to that which is obtained by recording the natural effects of lightning (SEA), and interference from outside sources will be kept to a minimum.

Building the receiver

The schematic diagram for the receiver is shown in Figure 12.1, and a complete parts list is provided in Table 12.1. The general layout of the unit and the simplicity of its construction can readily be seen in the photograph of the receiver with its cover removed (Figure 12.2). The small space that is required to set up the complete flare monitoring station (Figure 12.3) is an attractive feature of this project, which can provide the user with a considerable amount of scientifically useful information and many hours of enjoyment.

All of the receiver's parts are inexpensive and easy to obtain with the exception of two tuning coils, L1 and L2, which are no longer included in the Miller Company's catalogues. However, they may be purchased through the program by those who wish to build a receiving unit. Alternatively, Miller inductor coils #9006 can be substituted for the older models. In this case, a second capacitor with a rating of 470 picofarads should be added to each tuning coil by connecting it in parallel with the existing capacitor.

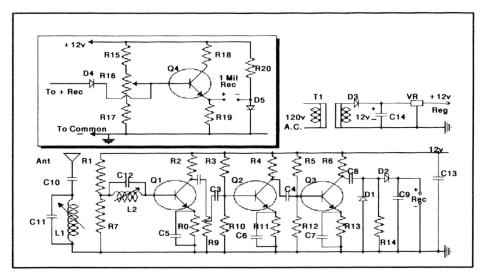

Figure 12.1 The schematic diagram for the very low frequency (VLF) receiver which is described in the text.

Circuit description

This version of very low frequency radio receiver is basically a tuned high-gain amplifier which is similar in concept to a circuit previously described by Carl Chernan (1980). It consists of a three-stage, high-gain transistor amplifier with a potentiometer gain control between the first and second transistors. For simplicity, the amplifier is capacitatively coupled throughout, with no interstage transformers.

The first stage has two tunable inductors (tuning coils) which are mounted parallel to each other on the front panel. Each inductor is shunted by a 0.001 microfarad capacitor to form an 'LC' tuned circuit. (An LC tuned circuit is one which is made up of a capacitor and inductor connected together. The product of the two values determines the resonant frequency, which in this situation is varied by adjusting the coil's core-screw.)

An antenna is coupled to the first inductor. The second inductor is connected to the base of the first transistor. The third transistor feeds a rectifier–integrator network which provides the direct current for a one hundred microamp strip-chart recording device.

The basic circuit can easily be modified so that it drives a one milliamp recorder rather than the typical microamp apparatus, by the addition of one additional transistor and a few resistors and diodes. These parts can simply be added to the original circuit board, and are listed separately at the bottom of Table 12.1. The twelve-volt positive and negative lines of the modification are connected to the same points on the original board. This change is shown schematically in the inset of Figure 12.1 which depicts the supplementary driver.

119

Table 12.1 *Parts list for VLF receiver*

Schematic letter code	Description	Radio Shack part number
C2, C3, C4, C6, C7	2.2 mfd	$272 - 1435$
C5	0.1 mfd	$272 - 135$
C10, C11, C12	0.001 mfd	$272 - 126$
C8	10.0 mfd	$272 - 1013$
C9	100.0 mfd	$272 - 1016$
C13, C14	50.0 mfd	$272 - 1015$
D1, D2	1N34 rectifier diodes	$276 - 1123$
D3	1N4001 rectifier diode	$276 - 1101$
L1, L2	Inductor coils – Miller #6319	(see text and footnote)
Q1, Q2, Q3,	NPN transistors 2N4401	$276 - 2058$
R1, R3, R5	100K ohms	$271 - 045$
R2, R4, R6, R7, R10, R12, R14	10K ohms	$271 - 034$
R8, R11, R13	470K ohms	$271 - 019$
R9	10K potentiometer	$271 - 1715$
T1	12-volt transformer	$273 - 1385$
VR	7812 12-volt voltage regulator	$276 - 1771$
Metal cabinet		$270 - 253$
BNC antenna cable connector (female)		$278 - 105$
Miniature audio plug and jack		$274 - 251 + 274 - 283$
Plug and jack for power supply		$274 - 1563 + 274 - 1569$
Knob for potentiometer		$274 - 402$
Perfboard to mount components		$276 - 1395$
1-centimeter spacers to mount board (4)		$276 - 195$

Parts List – supplementary driver for 1 milliamp recorder		
D4	1N34 rectifier diode	$276 - 1123$
D5	1N4001 rectifier diode	$276 - 1101$
Q4	NPN transistor 2N4401	$276 - 2058$
R15	68K ohms	$271 - 038 + 271 - 042$
R16	10K potentiometer	$271 - 1715$
R17	4.7K ohms	$271 - 030$
R18, R20	3.0K ohms	$271 - 028$
R19	470K ohms	$271 - 019$

Note: If Miller inductor coils 9006 are substituted for those listed in the table, an additional capacitor with a rating of 470 picofarads should be added to each tuning coil.

Figure 12.2 The receiver's simple design and recommended parts-layout is apparent in this photograph of the unit with its cover removed.

In either situation, the transformer is a twelve-volt, AC wall-plug type, with the twelve-volt line run to the power connector.

Construction

The transistors, resistors and capacitors are arranged on the twelve-by-seven-centimeter perfboard, in a manner which is similar to that shown in Figure 12.2. All of the power supply components, with the exception of the transformer, are placed along one edge of the perfboard. Three wire-leads are run from the completed board to the 10K potentiometer that is used to vary the receiver's sensitivity, and is mounted on the receiver's front panel. When soldering the transistors in place, be very careful not to overheat them. Use a heat-sink between the transistor and connection when soldering the joint.

The two inductors are also mounted on the front panel, on centers twenty-five millimeters apart with the inductor core-screws projecting outward for ease in tuning. This arrangement provides a mutually inductive coupling between the

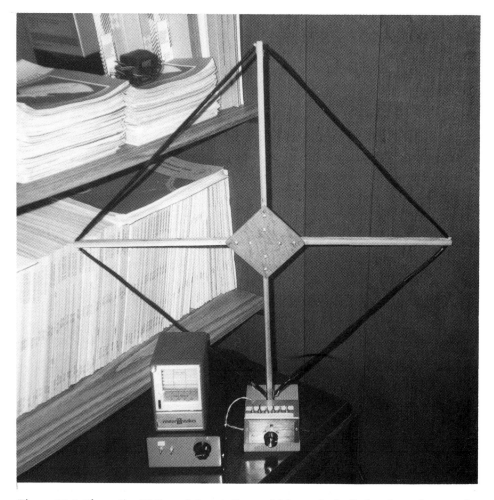

Figure 12.3 The entire VLF receiving station, which can be built for about the cost of a moderately sized reflecting telescope.

two coils to transfer the antenna signal to the base of the first transistor. The two 0.001 microfarad capacitors are mounted directly on each of the coil terminals.

The antenna connector, the mini audio-jack and the power connector are mounted on the back panel of the receiver's cabinet. After all of the components have been properly placed and appropriately connected, the assembled perfboard is secured to the bottom of the cabinet at each corner, using the four one-centimeter metal stand-off bushings. It is important that all of the connections are well-soldered. Mechanical connectors, with the exception of those listed in the parts list, should not be used.

Antennas

Antennas which are suitable for VLF reception generally fall into three main categories: long wire; vertical (such as a rain-gutter downspout or length of aluminum pipe), and loop-type antennas. The first two of these designs require that the receiver be solidly grounded to a steel or brass rod which has been placed near their base. (The rods can also be used for lightning protection.)

Either the long wire or the vertical antenna is connected to the receiver with a short length of co-axial cable. As a general rule, the shorter the cable's length, the better. Either the receiver's cabinet or the outside shield of the cable should be securely grounded, preferably to a three or four-meter ground-rod driven fully into the soil. A cold-water pipe or similar ground can be utilized in some circumstances, but the importance of an excellent ground for these antennas cannot be over-emphasized.

Any of several types of devices can be employed as a vertical antenna, and it may be wise to experiment with different designs in the event that the receiver's reception is not up to the expected level. The simplest type of vertical antenna is the downspout of a rain-gutter, and in many situations this will provide a completely adequate signal.

If the downspout does not provide a strong signal, a three-meter 'C-B whip,' or eight-meter length of aluminum tubing or pipe, three centimeters in diameter, can be employed. The latter should be erected vertically and suitably supported by non-conductive guy wires or a similar structure. A metal observatory dome will also provide a strong and reliable signal, but again the cable-shield or receiver must be grounded. Of course any of these devices must be insulated from the ground in some manner, perhaps by placing the antenna on a wooden sub-structure. The antenna should be located away from any existing power and electrical wiring, and the co-axial cable receiver-lead should be screwed or (preferably) soldered directly to it.

Building a tuned-loop antenna

A tuned-loop antenna can also be employed. Since this type of antenna is small, can be located indoors and eliminates the need for the receiver to be grounded, it provides an attractive alternative for observers who live in apartments or are otherwise unable to install an exterior device. In many cases, the signal that the loop provides will be equal or superior to that of the other types. However, its usefulness may vary according to the observer's location and their station selection.

The antenna loop consists of 125 turns of number 26 enamelled copper wire, placed in the notched ends of wooden cross-arms and secured by plastic ties (Figure 12.4). When building this framework, the cross-arms are first cut to a

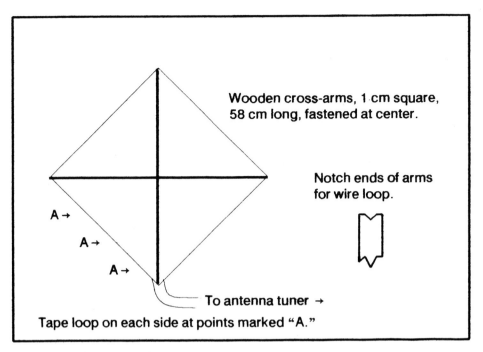

Wooden cross-arms, 1 cm square, 58 cm long, fastened at center.

Notch ends of arms for wire loop.

A →

A →

A →

To antenna tuner →

Tape loop on each side at points marked "A."

Figure 12.4 The antenna loop.

length of fifty-eight centimeters, then notched and attached at their center and braced by a small square plate of plywood sheet.

The antenna wire should then be carefully wound around the structure and each end secured separately on either side of the base, leaving a few centimeters of wire free to be attached to the antenna tuning system. After the wire has been wound around the arms, it should be bundled by electrical tape at three places per side, as indicated on the figure.

Since the loop antenna is very sensitive to the direction of the signal's source, the cross-arm structure should have a dowel, or short length of metal rod inserted into the bottom of its vertical arm. Then a suitable hole can be drilled into a wooden base and the rod inserted into it so that the antenna can be freely rotated, and once set will remain in a fixed position.

All switches and capacitors are mounted in a small plastic box according to the schematic diagram for the antenna tuning unit which is shown in Figure 12.5. The tuning box should be screwed or glued to the antenna base in a position that will allow the antenna to be rotated. Each capacitor's rating should be labeled on the outside of the box opposite to the appropriate switch, so that the values can be referred to during the tuning process. A list of the parts which are required to construct the antenna system is provided in Table 12.2.

Table 12.2 *Parts list for antenna*

Description	Radio Shack part number
Capacitors:	
50 pfd (1)	272 − 121
100 pfd (1)	272 − 123
200 pfd (1)	272 − 124
500 pfd (1)	272 − 125
0.001 mfd (2)	272 − 126
Plastic box	270 − 230
Slide or toggle switches (6)	275 − 327
3-meter length of co-axial cable	RG − 58
BNC cable connector (male)	278 − 103
245-meters of 26 enamelled copper wire	————
Wood for cross-arm structure	(see text)
Plastic electrical tape (1-centimeter wide)	————

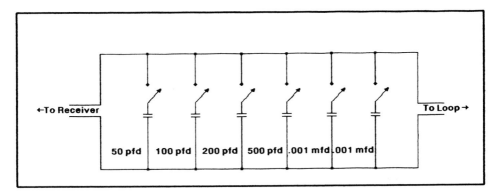

Figure 12.5 The schematic diagram for the loop antenna's tuning system.

Tuning the system

Tuning the receiver

Tuning to a specific frequency can best be accomplished by connecting an audio signal generator to the antenna input terminal through a 100K resistor, and an oscilloscope to the collector terminal on the final transistor. If one does not have access to this specialized equipment, it may be possible for a local radio-repair shop to tune the unit.

In those situations where this equipment is not available, the receiver can be tuned manually, by carefully adjusting the inductor screws according to the approximate measurements which are provided in Table 12.4. Then, with the

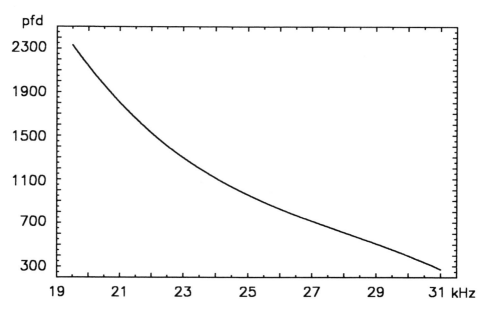

Figure 12.6 The loop antenna is tuned manually by using the switch settings which are indicated in this diagram and instructions from the text. Note that the settings are additive.

antenna and recorder connected, the signal should be maximized by slightly adjusting the tuning screws until the recorder-needle reaches its peak position. These final adjustments should be made *very slowly* since there is a long time-constant on the recorder circuit. Tuning information for several suitable VLF stations is listed in Table 12.3.

Tuning the loop antenna

Connect the antenna to the receiver and position it in a vertical plane. Switch the receiver on and turn up the sensitivity until the recorder needle is mid-range in its scale. The receiver's tuning coils should also be set as we have described above, or if a particular station has not been selected, at the mid-point of their range.

When choosing a station, select a strong one at some distance rather than a nearby source. (A very intense signal will overpower the receiver.) Start the tuning process by turning on various switch combinations until the recorder needle moves upward. Guides to the preliminary settings for many of the available frequencies are provided in Figure 12.6.

By trial and error, find the switch combination which results in the maximum signal, remembering that the switch values of capacitance are additive. After the maximum needle amplitude has been obtained, adjust the receiver's tuning coils so that the maximum signal strength is again at its peak. It is likely that the

Table 12.3 *Some very low frequency stations available for SID detection*

Frequency (KHz)	Call	Output	Location
20.76	ICV	500 kW	Tavolara, Italy
21.06	HWU	200 kW	Le Blanc, France
21.37	GYA	120 kW	London, England
21.40	NSS	1 MW	District of Columbia, USA
22.30	NAA	1 MW	Maine, USA
22.30	NWC	1 MW	Northwest Cape, Australia
22.30	NPC	1 MW	Washington, USA
23.40	NPM	1 MW	Hawaii, USA
24.00	NAK	1 MW	Maryland, USA
24.00	NBA	1 MW	Balboa, Canal Zone
24.00	NPM	1 MW	Hawaii, USA
24.80	NLK	1 MW	Washington, USA
25.30	NAA	1 MW	Maine, USA
25.80	NSS	1 MW	District of Columbia, USA
26.10	NPG	1 MW	California, USA
26.10	NPM	1 MW	Hawaii, USA
28.50	NAU	100 kW	Aquada, Puerto Rico
28.50	NPL	500 kW	California, USA
28.60	RAM	1 MW	Moscow, USSR

Reference: Chernan, 1978.

Table 12.4 *Approximate inductor settings*

Frequency (KHz)	L1 (mm)	L2 (mm)	Frequency (KHz)	L1 (mm)	L2 (mm)
21	11	9	22	16	12
23	17	13	24	18	15
25	19	17	26	20	17
27	20	18	28	21	19
29	21	20	30	22	21
31	22	21	32	23	22
33	24	23	34	25	23
35	25	24	36	25	25

receiver's sensitivity will need to be decreased slightly during and after this process, so that the recorder will remain on scale when a SID occurs. Unlike the other types of antenna, the loop is very directional; consequently, it should be rotated very slowly in the vertical plane until the maximum signal intensity is reached. A change in antenna angle of as little as ten degrees can cause a considerable change in the needle's amplitude.

The recorder

The recording device which is recommended for this project is the Rustrak Model 288. Of course recorders other than the Rustrak can be utilized for this purpose, but they should be of a similar technical specification, and geared so that they move the paper tape recording medium along at a speed of exactly twenty-five millimeters per hour. Each day's activity is transferred to the specially designed paper tape by impact, so that no ink or additional material is required. Some recorders operate on direct current, and in these situations a twelve-volt regulated power supply, similar to that which is used to power the receiver, is also employed with the recorder circuit.

Used recorders can often be purchased from universities where they are generally no longer in use, or from other sources such as large electronics companies which deal in end-lots and refurbished equipment. Either a microamp or milliamp recorder can be used, although the circuit will need to be modified as described above if it is to drive a milliamp recorder. If a new Rustrak recorder is to be obtained, it can be ordered from Rustrak Instrument Division, Gulton Industries Inc., Manchester, NH 03103, USA.

When a milliamp recorder is used, the 0.6 volt drop of diode 'D5' serves as a reference voltage to zero the recorder needle. To zero this recorder properly, turn the receiver's sensitivity potentiometer (counter-clockwise) to zero, and set the recorder to zero with potentiometer R16. Otherwise, the recorder is set to zero with the adjustment which is located on the front of its case. After this has been accomplished, the sensitivity of the receiver should be increased to about the mid-level position of its scale.

Timing the recording

Once the station is operating successfully, the recorder should be allowed to record an entire month's activity on a single, continuous roll of chart-paper. When beginning the recording, and at any interruption thereafter, the time marking on the tape should be carefully aligned with the zero line on the recorder, and the exact date and time (within one-minute) noted on the trace.

It is best if this process is carried out on the exact hour, and all times should be referred to the Universal Time format. The timing for each SID event is very important during the analysis of the tape-record, and consequently the observer should make every effort to ensure that the correct time is maintained by the recorder. Again, when any time or other adjustments are required, they should be noted directly on the tape.

In the preceding chapter we indicated the importance of data acquired in this manner to researchers at National Oceanic and Atmospheric Administration (NOAA). If the observer does not intend to analyze the record him or herself, the tape should be sent to the program (see address elsewhere in the preliminary pages of this book) as soon as possible after the end of each month, where the events will be determined.

Observers who wish to reduce their own data should first obtain from the author details of the information which is required by NOAA. Then, after the events for a month have been derived, the results should be sent to the program so that they too can be included in the monthly report.

I will be happy to answer any questions concerning these systems, and encourage those who are interested in this special type of solar observing to participate in this intriguing and productive endeavor.

The Edenvale magnetometer and recorder

The prototype for this simple, but very effective amateur magnetometer was designed by A. McWilliams of the Department of Physics and Astronomy, St Cloud State University, St Cloud, Minnesota, USA. The version which is presented here (Figure 12.7) and the recording device were developed by M. Daniel Overbeek, from the Republic of South Africa.

Although the data which are acquired with this instrument are not useful to professional astronomers, they clearly display the profound effect of the Sun's activity bursts upon the Earth's environment. Consequently, the project is a very attractive one for the amateur astronomer with both solar and electronic interests. The magnetometer is quite sensitive and therefore somewhat tedious to adjust initially, but its construction is inexpensive and straightforward, and a number of amateurs have built and operated them. Several types of motors, gears and other parts can be substituted for those which are indicated on the drawings, and so I have not provided a specific list of parts. However, the materials should be easy to obtain in most locations.

The magnetometer can be remotely situated and connected to the rest of the equipment by a three-core cable (Figure 12.8, conductors a, b and c). It should be covered in order to minimize exterior effects and stray light. The instrument can utilize a commercial rather than home-made recorder and a Rustrak model 288, one-hundred microamp unit is suggested for this purpose. Only the electrical circuitry which is shown at the top of Figure 12.8 is required in this instance. One recorder lead is connected to the point between the two light dependent resistors (LDRs) which is shown going to ground. Line 'C' is then disconnected, and the second wire to the recorder is connected to the slider on the zero adjustment through an adjustable resistor (50K potentiometer) that is inserted between the two units.

Figure 12.7 This sensitive magnetometer can be used to record disturbances of the geomagnetic field.

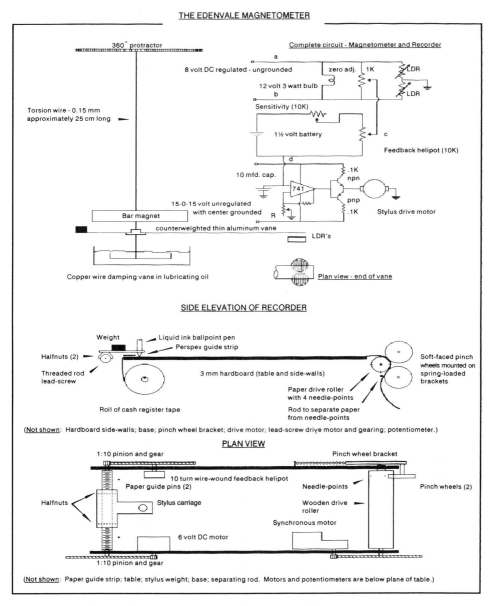

Figure 12.8 The simple construction of the Edenvale magnetometer and recording device is evident in this diagram. The magnetometer should be covered and placed in an area which is isolated from the recorder and local effects.

Construction

Several magnet suspension systems have been used with good success. These have included (in addition to the system that is shown on the drawing) suspensions which employed a twenty-five-centimeter length of nichrome wire 0.06

131

millimeter in diameter, and a slightly shorter 0.09 millimeter stainless steel version. A suitable magnet can be removed from a magnetic door latch; it should be about two and a half centimeters in length.

The upper end of the torsion wire is secured to the protractor with epoxy cement, or it can be similarly attached to a short length of six-millimeter tubing which is passed through a hole in the suspension arm and attached to a radio-style knob. When either unit is rotated, the torsion on the wire can be varied easily. The lower end of the torsion wire can also be secured with epoxy cement.

The LDRs are cadmium sulphide light-dependent resistors. The small twelve-volt dash-bulb is placed so that it is several centimeters above the sensors and lights each equally. It will be noted that the two halves of the zero adjustment potentiometer together with the LDRs form a Wheatstone bridge circuit. The Wheatstone bridge is slightly unbalanced, and has an output potential c to ground which is equal and opposite to the potential c to d. When these are not equal, the resulting potential at the input terminals of the amplifier causes the stylus-drive motor to run until a balance is achieved.

The resistor labeled R is made adjustable so that its value can be varied until the stylus-drive motor is neither too lively nor too sluggish. A fixed resistor can then be substituted. The transistors can be any npn/pnp pair which is capable of passing seventy milliamps or whatever the motor requires. Unit 741 can be any operational amplifier with characteristics which are roughly similar to the 741.

The Edenvale recorder's chart table is thirty-eight centimeters long and ten centimeters wide. Holes in the recorder's hardboard sides serve quite well as bearings for the lead-screw and paper drive roller. The only machining which is required is to reduce the ends of the lead-screw to form bearing shafts, and to provide a means of attaching a pinion and a gearwheel on the ends of the screw. A roll of cash-register tape will generally last for an entire year.

Adjustment

Start with an untwisted torsion wire and the magnet in the magnetic meridian. Rotate the protractor (or knob) slowly through an angle, alpha (α), which will usually vary upwards from 500 degrees. As the wire is twisted, the bar magnet will gradually be forced out of the meridian until it has rotated through nearly ninety degrees. When α reaches a *critical* value, the magnet will suddenly swing over to a new equilibrium position. The operational mode is just short of this value. In this state, the instrument senses changes in H, (the intensity of the horizontal component of the Earth's magnetic field), and not of D (the field's direction). Sensitivity depends upon angle α and is typically such that a change of 0.5 percent of H will cause the vane to swing through about one degree.

Set α to the required sensitivity by using a small magnet of known strength at a measured distance above or below the bar magnet. The aluminum vane must be able to rotate independently of the magnet so that it can be centered between the sensors after the magnet has assumed its operating position. Then adjust the vane relative to the magnet until it covers the LDRs fairly equally.

Set all potentiometers and the stylus carriage to the center of their travel and switch the amplifier on. The stylus should move. Bring it back to center by using the zero adjustment. (If the stylus runs away, the feedback sense needs to be changed from positive to negative by reversing the motor leads.) Change the settings of R and the sensitivity controls as necessary. The zero adjustment may also need to be varied during the tuning process.

The recorder can also be coupled with a VLF receiver and used to monitor the atmosphere for sudden ionospheric disturbances (SIDs). However, the timing of SID events is critical, and if the recorder is to be put to this use conventional chart-paper should be employed in place of cash-register tape. Each foot of this recording medium is equally divided into hourly intervals which are sub-divided into fifteen-minute segments so that the exact time can be conveniently set and maintained.

13

Computation of observer statistical factors and other useful equations

When Rudolf Wolf originated the relative sunspot number he defined it as an estimate, rather than a precise measurement of the Sun's activity. Each observer derives this quantity in the same manner, but each is also subjected to different average weather conditions, uses instrumentation which can be quite dissimilar to that employed by others, and counts the groups and spots as they alone interpret their structure. Understandably, the values which result from this process frequently vary by a wide margin.

In his description of the basic reduction techniques for the American Relative Sunspot Number (Shapley, 1949), Alan Shapley discusses estimates for one particular day which range from forty-one to over 300! While it is true that these particular observations were made when the program first began and its collaborators were new to sunspot counting, similar discrepancies can be found among the contributors to the Zurich sunspot numbers when their data are examined over time.

For example, when observations obtained at the Greenwich Observatory are compared with those computed by Zurich for the interval 1921–41, the numerical ratio between the two sets of data varies by as much as fifty percent. In order to effectively equalize a difference of this kind, it is necessary to scale each collaborator's estimates through the application of an observatory scaling constant known as a K-factor. Shapley believed that it was also necessary to apply an additional statistical component, W, to each contributor's estimated sunspot count.

The first of these constants, K, is intended to compensate (as much as possible) for differences in the instrumentation, average local seeing conditions and personal judgement among observers. Since the standard on which this constant is based is the historical series of relative sunspot numbers, the K-factor also fulfills the requirement of reduction to the same scale as the original Wolf series.

New sunspot observers are often curious about their K-factors, and seek a simple way to determine them. Although the method is not precisely accurate, an observer can acquire a good estimate of their scaling factor by simply dividing their daily estimates into the appropriate values of final sunspot numbers, and averaging the results. A realistic appraisal of the K-factor will require that at least fifty estimates are compared in this manner. Novices should remember that as

134

they become more experienced in the art of spot counting, their factor is likely to decrease since they will see more of the fainter spots.

When computing the American sunspot number, K-statistics are calculated for each observing station from at least a hundred observations. These estimates must have been obtained during adequate seeing conditions (i.e., when seeing is rated as either good or excellent) and when monthly-mean sunspot numbers have regularly exceeded 100. This second requirement is especially important since K-factors which are computed when sunspot numbers are low are generally significantly higher than those which are calculated near the maximum phase of the spot cycle.

If left unchecked this anomaly will almost certainly result in artificially high sunspot numbers for most of a cycle. In fact this difficulty was experienced during the early years of the American program, even though Shapley employed the set of final Zurich Relative Sunspot Numbers as a basis for deriving the initial K-statistics for his contributors. Over the course of several years, the new American values gradually exceeded those derived at Swiss Federal Observatory by greater and greater amounts, until they had diverged by fifteen or more units. Shapley attempted to correct this problem by introducing the concept of weighted estimates into the reduction process, but it is more likely that an eventual re-evaluation of each observatory's scaling factor was the real cure.

In spite of these constraints, new contributors to the program who are diligent in their efforts may have their observations included in the analyses of American Relative Sunspot Numbers after only a short time. However, their data are necessarily restricted through an arbitrarily imposed, lower statistical weight until such time as the monthly-mean spot numbers have returned to the target range, and new K-factors can be derived accordingly.

The computation of the K-statistic for use in the reduction of the American sunspot numbers is slightly more complicated than the simple approach which is suggested for curious newcomers. It is accomplished through the application of the following relation (Taylor, 1985):

$$a_i = \log K_i = 1/N \left(\sum_{j=1}^{N} \log R_{sj} - \sum_{j=1}^{N} \log R_{ij} \right).$$

When solving this equation, R_s represents a set of reduced daily values used as a standard for comparison: in our case, the set of final American Relative Sunspot Numbers for the computational period. The R_i component-series is composed of a matching set of data from a selected observer in the form dictated by Wolf ($R = 10g + s$), while N represents the number of data pairs which are used in the study.

An observer's estimates are not included in the computations when no sunspots are seen (i.e., on those days when R is zero). Of course, a sunspot number of zero is entirely possible when the Sun is undergoing a low activity

135

level. However, observations of the Sun's other phenomena during these times have shown that the Sun's overall activity still varies to some extent, and consequently these estimates should probably not be employed in the determination of scaling factors.

To be useful statistically each contributor's *K*-statistic should fall within the range, 0.50 to 1.50, although for technical and, to some extent physiological, reasons values skewed towards the *lower* end of the scale are preferred. To achieve this, all observers must be alert to the presence of the smaller and fainter groups of spots, since a failure to *see* these clusters is a primary cause of high observer *K*-statistics. One additional requirement is imposed on all regular observers: that is, their *K*-factor should not vary by more than ten percent between annual re-computations. In this way, a consistently high quality of data can be maintained.

The second constant, *W*, is a weighting-factor. In fact, *W* is a measurement of how well observers' *K*-statistics correct their unreduced data to the standard; a way to rank the *consistency* of the observer's estimates. To my knowledge, this procedure was never employed in the reductions at Swiss Federal Observatory, and has not been adopted in the International program which is currently administered by the Royal Observatory of Belgium. The weighting-factor which is utilized in the reductions of American sunspot numbers is computed according to

$$W_i = \frac{N-1}{\sum_{j=1}^{N} (\log R_{sj} - \log R_{ij})^2 - N_{ai}^2}.$$

According to Shapley, the regression analysis described by these two relations is an example of the type defined by Wald (1940) wherein both sets of variables (R_s and R_i) may contain discrepancies. A logarithmic approach was selected for these reductions in order to facilitate a more homogeneous data set, although the differences between the results which are achieved with this technique and a traditional method are actually quite small.

The Wald approach assumes that any variance within the data is equally distributed, and that only uncorrelated errors exist among the individual series. In general these requirements are likely to be fulfilled, although it is possible for the errors of two (or more) observers who are located in nearby areas to be correlated on some days, since both could experience similar weather conditions.

After the scaling and weighting-factors have been computed for each contributor, the relative sunspot number, R_a, for each day of the month can be computed through use of

$$R_a = \frac{\sum\limits_{i=1}^{N} W_i\, K_i\, R_i}{\sum\limits_{i=1}^{N} W_i}$$

and the result rounded to the closest integer for each day under consideration. The number of observers for any one day varies of course, but generally exceeds thirty for the computations of final American sunspot numbers. Monthly-mean sunspot numbers are simple averages of R_a.

Because the minimum estimate for any sunspot group is eleven, I am frequently asked how it is possible for the daily sunspot number to be less than this, but yet not zero. The reasons are quite simple. The scaling process itself can lower the numerical value of each estimate, and counts of zero and those of short-lived sunspots are averaged during the reductions.

A set of at least thirteen monthly-mean sunspot numbers is necessary for the computation of the smoothed-mean relative sunspot number, R_{sm}. This important statitistic is usually calculated according to a technique which is suggested by Waldmeier (1961), or

$$R_{sm} = 1/24 \left[N_{i-6} + N_{i+6} + 2 \left(\sum_{-5}^{+5} N_i \right) \right].$$

In this method N represents the final mean for the month under analysis, and the other mean values are taken accordingly. Thus the smoothed monthly-mean sunspot number always lags six months behind the more recent monthly-mean.

Final values of American Relative Sunspot Numbers for the interval 1944–1988 are provided in Taylor (1987, 1988b, 1989b, 1990). Annual means for the years before 1950 were averaged monthly-means; thereafter, all yearly means have been computed by averaging the daily values.

Some other useful equations

It is often convenient to know the *Carrington Rotation Number* (CRN); the usual method for numbering the rotations of the Sun. The initial rotation of the Carrington system began on 9 November, 1853 (Carrington, 1863) and, according to the *Astronomical Almanac*, rotation number 1811 began 9.17 January 1989 (Julian Date 2 447 535.67). The *Julian Day* begins at Greenwich Noon and is numbered consecutively beginning with 1 January, 4713 BC. For example, Julian Day 2 447 528.0 is equivalent to a Universal Time of 1.5 January, 1989.

The period of the Sun's mean synodic (apparent) rotation is generally

137

assumed to be 27.2753 days, during which time the value for the Sun's central meridian decreases by 360 degrees. With these factors in hand, the rotation number for any date may be computed according to

$$\text{CRN} = \text{INT}\left[1811 + \left(\frac{\text{JD} - 2447535.67}{27.2753}\right)\right],$$

where JD represents the appropriate Julian Date, which can be obtained from the *Astronomical Almanac*, and a number of astronomical handbooks.

Since the Sun rotates differentially (its polar regions complete one full rotation in about thirty days – about five days more than its equatorial zones), I am occasionally asked for a simple method to determine the rate for a particular zone. Although this is a complex question, one convenient way is to apply the straightforward method outlined in Zirin (1988),

$$\Omega = 14.42 - 2.30 \sin^2 \theta - 1.62 \sin^4 \theta \text{ degrees per day,}$$

where Ω is the sidereal rotation rate and θ is the latitude.

Many students of the Sun find that it is interesting (and somewhat sobering!) to determine the approximate diameter of a large spot, or perhaps the length of a particular group. If the spot is located near the center of the Sun's disk and the observer has obtained a eyepiece equipped with a cross-line reticle such as we have described previously, these qualities can be determined according to the following procedure.

First, turn the eyepiece/reticle combination in its holder so that the group of spots moves parallel to one of the crossed lines. Now record the time which elapses between the instant that the leading edge of the group touches the opposite cross-line and the moment when the last of the cluster passes the line. Call the interval T. The group's length can then be determined by solving the following relation from Roth (1975):

$$\text{length (kilometers)} = 10\,855\ T \cos \delta.$$

In this equation, 'δ' represents the declination of the Sun on the date of observation. Tables of the Sun's declination for each day are also provided in the *Astronomical Almanac*.

Finally, professional astronomers measure the areas of sunspot groups in units of millionths of the solar hemisphere, and the clusters are frequently described in the literature in this manner. A convenient way to state the area encompassed by a large sunspot complex is to convert these units into a more familiar measurement, such as square kilometers:

1000 millionths solar hemisphere = 3036 million square kilometers.

For comparison, the surface area of the Earth is a little over 510 million square kilometers. Although it is difficult to conceive of these vast areas, the largest sunspot groups on record have encompassed more than ten thousand million square kilometers of the Sun's surface[1], while those with areas of a thousand million square kilometers are actually fairly commonplace! A simple method for deriving the area of a particular sunspot cluster is outlined in Chapter 9.

[1] The largest sunspot complex on record occurred during April 1947. Greenwich Observatory measured its area to be nearly sixteen thousand million square kilometers (Taylor, 1989c).

14

Solar eclipses and the amateur astronomer

Every so often a few of us are fortunate enough to witness one of nature's most spectacular events; the total eclipse of the Sun. Even though very special conditions must occur for them to take place, no other celestial display can compare with the astounding beauty and drama of a total solar eclipse. However, in spite of their impressive appearance, the scientific study of the eclipse phenomenon did not really get under way until the seventeenth century, and failed to add much to what was known about the Sun until over 200 years later.

Since eclipses were poorly understood by the majority of the world's people for many centuries, they have often been regarded with suspicion and mistrust. At times these fears were exploited by the old historians of the past, who occasionally manipulated the date of an eclipse, or even fabricated the events in order to influence the course of history in some way. According to astronomer J. B. Zirker (1984), the apparently high incidence of antique eclipses during times of great national change may well be related to these, rather than natural circumstances. As a result it is often difficult to determine the dates (and, in some situations, the reality) of the earliest events.

Nevertheless, the first reliable account of an eclipse is thought to have been recorded by the ancient Chinese sometime between 2165 and 1948 BC (the most likely year is 2137 BC) during the time of the Hsia dynasty. In his 1949 treatise, *Our Sun*, D. H. Menzel suggests that this eclipse may have led to the decapitation of the royal astronomers Hsi and Ho. According to the story, the two astronomers were consumed by wine and failed to perform the necessary rituals to drive off the dragon which was consuming the Sun, causing fear and chaos. Unfortunately, the original record of these events was destroyed around 223 BC and was not reconstructed until the fourth century AD, so the fate of Hsi and Ho remains uncertain. (However, astronomer S. A. Mitchell has noted that in light of this warning, no astronomer has since repeated the mistake of Hsi and Ho!)

When they can be adequately verified, eclipses are a valuable tool for historians as well as for astronomers. The eclipse which took place on 15 June, 763 BC has played an important role in correlating early Assyrian dates with the modern calendar. Similarly, Ptolemy's great work *Almagest* includes a list of the kings of ancient Babylon, Assyria and Persia, which, when coupled with the

140

eclipse record, has been useful in fixing the dates of eastern chronology (Mitchell, 1923).

The first solar eclipse to be predicted through the analysis of the historical eclipse record may have been the famous event which took place during 585 BC. The account states that the sudden onset of darkness so startled the Medes and Lydians that they ended their five-year war and sealed the peace with a double marriage (Menzel, 1949). Nonetheless, this eclipse is more memorable because it is thought by some scholars to have been foreseen by the Greek scientist, Thales of Miletus, after an examination of the early Babylonian eclipse record. As a result, the eclipse has gained considerable notoriety, although Thales' limited knowledge of astronomy has led some historians to question his ability to predict such events (Zirker, 1984).

Even though the ancient Chinese were the first to record an eclipse, the early Greeks were almost certainly the first to understand why they occur, and capitalize on them for scientific purposes. Moreover, many of the greatest achievements of the Babylonian astronomers transpired after the fall of Babylon in 539 BC and the Greeks had invaded the valley of the Euphrates; a series of events which may well have influenced Babylonian thinking (Mitchell, 1923).

One of the earliest scientific applications to utilize measurements obtained during an eclipse was conceived by the great Greek astronomer Hipparchus, for his estimate of the separation between the Moon and Earth. In his experiment Hipparchus coupled two observations from the eclipse of 130 BC with the first use of astronomical trigonometry, to derive a distance ranging from sixty-two to seventy-four Earth radii between the two bodies. This is certainly a startling result when one considers that the modern value averages a little over sixty radii!

More recently, eclipses have played an important part in our understanding of the Sun, since before the invention of the coronograph and other sophisticated equipment they offered astronomers their only opportunity to discover the secrets of the prominences, corona and chromosphere. In addition, observations at eclipse have been important in astronometric research, in the behavioral studies of birds, mammals and lesser life-forms, and in the invention of other instruments such as the spectrohelioscope, to cite just a few examples.

Unfortunately, eclipses have continued to be regarded with a certain amount of confusion and contradiction, especially among those who live in the world's undeveloped areas. Anthropologists must be particularly intrigued by the paradoxical nature of the 1922 Australian eclipse. On the one hand, the event was used by Campbell and Trumpler to verify Einstein's prediction of the displacement of starlight in the field surrounding the Sun. But at the same time, near the site where these observations were gathered, the aborigines believed the astronomers were trying to capture the Sun in a net! (Menzel, 1949).

Sadder still are the modern-day instances of ignorance and apathy with which many of those who attend eclipses are familiar. It is impossible to fathom the

141

motives of car drivers who casually turn on their headlights and continue on their way during totality, or those who warn schoolchildren and others to avert their eyes from the eclipse rather than teach them the simple ways to enjoy nature's greatest spectacle with complete safety.

Of course the conditions which produce an eclipse are now well understood, and astronomers can predict their occurrence for almost any time in the future. We know that the greatest number of solar eclipses which can take place in any one year is five, while the fewest is two, so that during many years more eclipses of the Sun are possible than those of the Moon; a fact which seems surprising to many.

About thirty-five percent of these eclipses are partial, thirty-two percent are annular, and twenty-eight percent are total. The remainder is made up of events which are combinations of annular and total. They are brought about under special circumstances, such as an extraordinary distance between the Moon and Earth, or rarer still, when the dark lunar shadow contacts the Earth at an extreme angle (Oppolzer, 1962; Meeus *et al.*, 1966).

In order for a *total eclipse* to take place, several conditions must be fulfilled. First of all, the Moon must be in a location which results in the darkest part of its shadow falling on some part of the Earth.

The Moon's shadow is actually composed of two parts: an outer shadow which grows wider with distance and encompasses a large portion of the daylight hemisphere during a typical eclipse, and a dark, narrow inner cone. The effect of these shadows on the Earth is shown in Figures 14.1 and 14.2. An individual in the outer, or *penumbral shadow* sees only a *partial eclipse*. Since the

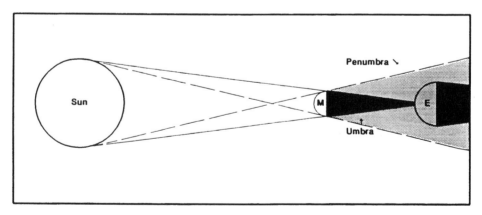

Figure 14.1 The Moon's shadow during a solar eclipse is actually composed of two parts. An observer on Earth sees a total eclipse only when he or she is located within the path of the umbral shadow. The penumbral shadow does not cover the entire daylight hemisphere as appears in the diagram, but rather extends over a wide band centered upon the path of the umbral shadow.

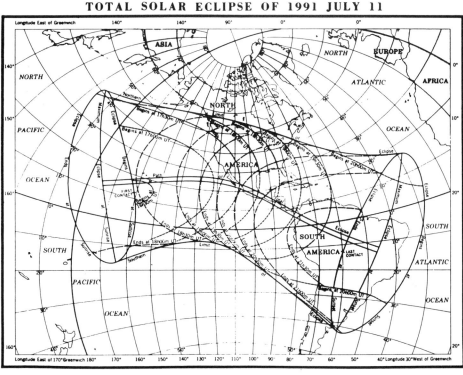

TOTAL SOLAR ECLIPSE OF 1991 JULY 11

Figure 14.2 The path of the Moon's inner and outer shadows during the July 1991 eclipse. This diagram was prepared by William Harris of the US Naval Observatory in Washington, DC.

remaining portion of the Sun's light is so intense, those who have not been forewarned of the event are often unaware that an eclipse is in progress.

Alternatively, a viewer within the inner, *umbral shadow* experiences either a total or annular eclipse. The length of the umbra varies only slightly, ranging between about 367 000 and 380 000 kilometers. However, the Moon's elliptical orbit causes the distance between the two bodies to vary by a much larger amount, so that it falls somewhere between 349 000 and 401 000 kilometers (Zirker, 1984).

The Moon must be near its closest approach to the Earth (*perigee*) for the eclipse to be total. Only then is its angular diameter large enough to completely block the Sun, which is about 400 times the Moon's diameter. On the other hand, when the Moon is near *apogee* the darkest portion of the Moon's shadow does not reach to the Earth's surface, or does so only momentarily, and an *annular* or *annular–total–annular* eclipse takes place.

During an annular eclipse, a bright ring of solar surface continues to surround the Moon at eclipse maximum, and many of the spectacular phenomena which

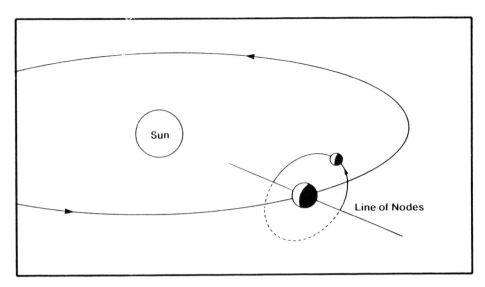

Figure 14.3 The lunar orbit intersects the ecliptic at two places called nodes. A line extending through these points must point directly at the Sun for an eclipse to take place.

appear during a total eclipse are obscured by the remaining photosphere. When the more unusual annular–total–annular eclipse occurs, the umbral shadow just touches the Earth for a brief interval, and the geographical area which experiences the few seconds of totality is extremely limited.

If the Moon's elliptical orbit were in the same plane as the Earth's path around the Sun (the ecliptic) a solar eclipse would take place every month. However, the lunar orbit is inclined from the ecliptic by about five degrees. These orbital planes intersect twice each month, at places which are known as 'nodes.' An imaginary line drawn through these points is called the *line of nodes*, and a total eclipse can take place only when the Moon is nearly new and the line of nodes points directly towards the Sun. The basic geometry of this relationship is shown in Figure 14.3.

Since these circumstances occur when the Sun appears to be at opposite locations of the celestial sphere, the eclipse 'season' repeats at intervals of about six months, and an eclipse is then *inevitable* at each node (Zirker, 1984). There is some latitude in the required placement of the three bodies which will still result in an eclipse, and so more than two events are possible each year. On occasion, when the geometry is just right, two eclipses can take place near each node. Moreover, when the first eclipse of the calendar year occurs in January, a fifth event can also take place (thirteen new moons occur during such a year). However, this happens only rarely: the next occurrence is not scheduled until the year 2160!

One other variation in the geometry of eclipses should be considered. That is, the place of the Moon's orbit does not remain at the same position of the ecliptic but gradually shifts its location westward, or *regresses* along it, completing one full revolution in slightly over eighteen and a half years. Since the nodes move counter to the orbital path, the Sun crosses the same node of the Moon's orbit in a little less than one year (346.62 days), an interval which is designated as an *eclipse year*.

The eclipse year is thought to have been defined for the first time around 300 BC by the Chaldean astronomer Kidinnu, after studying the long eclipse record begun by the Babylonians centuries beforehand (Batten, 1971). The ancient Babylonian astronomers kept track of time by observing the movements of the Sun and Moon, believing their movements to be symbols of law and order. Since they also thought that these motions controlled human events, they measured them as carefully as their methods would allow. Eventually, the accuracy of these data improved considerably, and a cyclic pattern appeared whereby eclipses were found to repeat themselves in a little over eighteen years.

The discovery of these cycles, known as the *Saros* (the Greek word for 'repetition') is often credited to the Babylonians, or occasionally to Edmond Halley who described and named them in the seventeenth century (Todd, 1899). There seems to be little doubt that the Babylonians, if nothing else, compiled much of the eclipse record on which the cycle was originally based. However, while some scholars believe that the ancient Babylonians could predict total solar eclipses through use of the Saros, most feel it is more likely that their knowledge limited them to the prediction of lunar eclipses. Thus the evidence appears to favor the early Greeks for the actual discovery of the cycle.

What are the conditions which yield the Saros? A lunar cycle (the period of time between new moons, or *synodic month*) is a little over 29.53 days in length. As a result, the durations of 223 lunar cycles and nineteen eclipse years are very nearly the same (6585.32 and 6585.78 days, respectively). Thus the circumstances which produce an eclipse are closely duplicated in eighteen years ten and a third, or eleven and a third days, depending upon the number of intervening leap-years.

Consecutive eclipses within a Saros series are similar in type and duration because the Sun and Moon have then returned to their original positions at the node, with the Moon close to its previous distance from the Earth. Successive eclipses do not reoccur at the same location as their immediate predecessors, however, since during the extra one-third of a day of the Saros cycle the Earth rotates through an additional 120 degrees of longitude, causing the path of the following eclipse to fall to the west. As it turns out, the eclipse which repeats near the longitude of the first event actually occurs three Saros cycles (about fifty-four years, one month) afterwards.

To illustrate these features, a list of twentieth century eclipses which are members of Saros series number 136 is provided in Table 14.1.

During the time it takes to complete one Saros cycle an average of forty-two eclipses take place. All but one of these events are related to additional Saros

Table 14.1

Date	Type	Mid-eclipse		Maximum duration (minutes)
		Latitude	Longitude	
17 May 1901	Total	$-02°$	$-97°$	6+
29 May 1919	Total	$+04°$	$+17°$	6+
8 June 1937	Total	$+10°$	$+130°$	7+
20 June 1955	Total	$+15°$	$-117°$	7+
30 June 1973	Total	$+19°$	$-06°$	7+
11 July 1991	Total	$+22°$	$+105°$	6+

series running in parallel, each of which is given a number. The numbering scheme for each group of cycles was devised by G. Van Den Bergh in 1955 (Van Den Bergh *et al.*, 1955). According to Van Den Bergh's scenario, odd numbered cycles are those which take place at the Moon's ascending node (when the Moon crosses the node while moving northward) and those assigned even numbers occur at the descending node.

Each Saros series begins with an eclipse which takes place at a high latitude. Succeeding members of the series gradually wind their way towards the equator, appearing first as partial, then as annular or total eclipses at mid-latitudes (Table 14.1). Subsequently, the eclipse series moves towards the opposite pole, and again becomes partial. For practical purposes, the group of cycles ends as the last partial eclipse occurs near the pole, so that the series does not extend indefinitely (Menzel, 1949).

On the average, a Saros series contains seventy-three eclipses and therefore lasts for some 1315 years. Other cycles appear in the eclipse record, and at least one of them extends over a much longer period than the Saros: the *Inex* series. The Inex is similar to the Saros in that it too results from two natural cycles of nearly the same length. This time, however, the cycles are the synodic month and the *draconic month* (the Moon's period with respect to a node, which is approximately 27.21 days).

Since 358 synodic months equals 388.5011 draconic months, an eclipse which belongs to the Inex series repeats at the *opposite* node almost twenty-nine years (358 new moons) afterward, and there is a tendency for the eclipses to alternate between annular and total. About seventy Inex series run in parallel at any one time. Each contains about 780 eclipses, and consequently a series lasts for 23 000 years! Van Den Bergh found that every eclipse could be related to a specific Saros series, and that every Saros was represented within the Inex series. Very-long-range predictions frequently utilize the Inex, rather than Saros record, even though it does not necessarily give a reliable indication of the eclipse type or location.

Generally only a relatively small number of people are fortunate enough to

Figure 14.4 Typically, mid-eclipse is seen as it appears in this photograph taken by Sky & Telescope *Associate Editor, Dennis di Cicco, during the October 1977 event. The original photograph shows coronal streamers to more than four solar radii. Photograph courtesy of Dennis di Cicco and* Sky & Telescope *magazine.*

live within the path of total eclipse because the umbral shadow is never very wide, and frequently spends much of its time over water. Since total eclipses take place (on the average) every one and a half years, the probability of occurrence at any one place is only about once in 450 years. The long-duration eclipse of 11 July, 1991 was an exception as far as the number of potential viewers. Its path extended over several large metropolitan areas, and as a result the event's total phase could have been witnessed by the largest concentration of people in all of recorded history (Rao, 1988). As a general rule, however, most of us must travel long distances in order to view these awe-inspring events.

When observing the partial phases of an eclipse telescopically, a solar filter such as that employed for normal observations is **mandatory**. The same precautions which apply to viewing the Sun under non-eclipse conditions **must** be followed when observing the partial phases, and for the entire annular eclipse. For non-instrumental viewing, shade fourteen welder's glass, or the solar filter held up to the eye are choices which offer good eye-protection. No filter is needed (or wanted!) during the *maximum* phase of a total eclipse such as that shown in Figure 14.4, but the observer must not continue to view the Sun without protection after 'Baily's Beads' reappear (see below).

It is frequently convenient to divide an eclipse into specific stages, each of which can be accurately timed. The timing and other observation of these phases, or 'contacts,' at various locations within and along the path of totality can prove to be an interesting exercise. Moreover, it is one of the few remaining areas where an amateur astronomer can participate in a scientifically meaningful eclipse endeavor.

For observers in the Northern Hemisphere, a typical eclipse takes place in the southern sky; so the Moon, moving thirteen times faster against the background of distant stars, approaches the Sun from the west. The first stage, the fleeting moment known as *first contact*, occurs just as the highest point on the leading lunar limb makes its initial appearance before the Sun; it is the instant of apparent external tangency between the two bodies.

Just prior to the onset of totality the dark umbral shadow appears on the horizon, looking as if it were a huge funnel-shaped storm. The shadow rushes across the landscape at a speed which is at least 1700 kilometers per hour, but can be five or more times this amount at high latitudes (Zirker, 1984). Occasionally, rapid changes of color are noted just before the shadow's edge passes, but it is more likely that the light will appear harsh in comparison with the atmospherically reddened light that is normally associated with dusk. Then the sky darkens and the phenomenon known as *Baily's Beads* appears along the lunar limb, as a few remaining bright points of sunlight thread their way through the Moon's rugged terrain.

The beads are named for the astronomer Francis Baily, who described them as 'a row of lucid points, like a string of bright beads ...' in 1836 (Chambers, 1890). However, a number of astronomers referred to the effect prior to Baily's description, and one of the first of these was Edmond Halley who viewed the phenomenon at the eclipse of 1715 (Todd, 1899). When the eclipse is annular but very nearly total, or annular–total–annular, the beads appear to form a 'necklace' which can be seen all around the limb. For obvious reasons, this phenomenon generally persists for a longer period of time near the edge of the total eclipse path. A related effect, the 'diamond-ring' occurs when one bright bead shines through a particularly large lunar valley and dominates all others with its brilliance.

At *second contact* true totality begins. For those situated near the center-line of the path of the eclipse the beads disappear and are replaced by the Sun's chromosphere. The intense light of the photosphere is now completely blocked by the black lunar disk, and the bright reddish ring of inner atmosphere caused by strong hydrogen-alpha emission can be seen. If any prominences are present they too appear, sprouting from the limb into space for distances which are often measured in the tens of thousands of kilometers.

As the eclipse continues, the Sun's extended atmosphere, the corona, comes into view. Since the combined light of the corona and scattered light from out of the path of totality is about as bright as the full moon, the sky seldom becomes as

148

Figure 14.5 This spectacular photograph of the June 1973 eclipse was taken by a five-man team from the High Altitude Observatory, the National Center for Atmospheric Research. The group utilized a special camera devised by G. A. Newkirk, Jr, equipped with a radial-density filter which dims the Sun's brightness at the limb far more than that several solar radii away. The photograph was supplied by Stuart J. Goldman of Sky & Telescope *magazine.*

dark as it is during a typical night. Therefore, stars as faint as third magnitude only appear rarely (Silverman and Mullen, 1973). According to these authors, those most likely to be reported are zeta Tauri, beta Canis Minoris and delta Crucis. Depending upon their altitude above the horizon, their brighter siblings, Capella, Procyon, Sirius, Pollux and Castor, are frequently sighted, as well as the planets Mercury, Venus, Mars, Saturn and Jupiter.

Since shortly after Schwabe published his announcement of a periodicity in the numbers of sunspots, the shape of the corona has been known to vary according to the phase of the cycle. At sunspot minimum, the corona appears to be elongated, with broad equatorial *streamers* and delicate *plumes* which extend from the Sun's polar regions (Figure 14.5).

Near the maximum of the spot cycle, however, the corona is more circular, and both narrow and broad *rays* reach outward in an irregular pattern. As is the case with most solar phenomena, these effects are related to magnetic activity on the Sun. Thus the shape and direction of the streamers, rays and plumes are all influenced by magnetic fields which originate within the Sun (Giovanelli, 1984).

The pale ghost-like appearance of the Sun's outer atmosphere is produced by sunlight which is reflected off free electrons in the corona out to a distance of about one-half of the Sun's radius, and as the light is scattered off dust particles in the solar system beyond that (Noyes, 1982). Visual observers usually report that the corona can be seen to extend for a distance of a few solar radii from the Sun, although the historical record contains reports which claim to have viewed the phenomenon for twenty or more radii (Chambers, 1890; Todd, 1899). Recent studies conducted aboard high-altitude aircraft have shown that the corona does extend to distances which are at least as great as nine million kilometers from the Sun.

The total phase of any eclipse is all too short; always less than seven minutes forty seconds (Zirker, 1984). Totality lasts longest at the equator, because at low latitudes the counteracting component due to the Earth's diurnal rotation is greatest. (The shadow's true speed is in excess of 3300 kilometers per hour. However, the Earth's rotational speed at the equator is about 1670 kilometers per hour in the direction of the shadow's path, and this slows its passage relative to an equatorial observer.)

The chromosphere stages a brief return as the corona fades from view, and at *third contact* totality ends as suddenly as it came. This contact is signaled by the reappearance of Baily's Beads along the trailing limb of the moon, and is followed by the second phase of partial eclipse. Eventually the last tiny vestige of lunar disk disappears from the Sun's eastern limb, marking *fourth contact*, the most elusive and difficult of the four contact timings.

It is a marvelous visual experience, and consequently I believe that amateur astronomers can best enjoy total eclipses by simply watching them. Those who wish to participate more actively could search for, and time the elusive 'shadow-band' phenomenon which results from light refracted in the Earth's atmosphere, or perhaps measure the speed of the Moon's shadow itself. On the other hand, observers with an electronic interest might devise an experiment that would utilize the equipment which is described in Chapter 12 to record any ionospheric effects that result from the changes in sunlight.

Others who are interested in the biological consequences of the sudden onslaught of darkness, could note the differing effects of the eclipse upon the local insect and animal population. (Both Polish and Indian scientists have determined that mammals are much less affected than invertebrates.)

Some excellent advice on photographing the various stages of an eclipse can be obtained from the Eastman Kodak Company (Rochester, NY 14650, USA). Request Publication No. P-150, *Astrophotograpy Basics*. Intriguing projects for the amateur astronomer are often described in magazines which are intended for the popular astronomy market. Those who are interested in these aspects are urged to seek out the pertinent articles. (*Sky & Telescope* magazine is a particularly good resource for projects such as these.) The last word in technical information about an eclipse can be obtained from the US Naval Observatory (*Eclipse Circulars*, Washington, DC 20392-5100, USA).

As has been mentioned, specific observations made within, and along the path of totality can be useful to the professional scientific community. However, these observations, while simple, must be made within the guidance of a structured program in order to have value. I would advise those who would like to take part in this type of activity to contact the International Occultation Timers Association (6 North White Oak Lane, St Charles, IL 60174, USA) as far in advance of the eclipse as possible.

Alternatively, data which are obtained by experienced amateur astronomers according to specific guidelines can be sent to the Nautical Almanac Office of the US Naval Observatory. Precise observations of the second and third contact times from locations along the path, particularly at its northern and southern limits, are needed. A report of this nature *must* pin-point the observer's site to within 50 feet of a prominent and identifiable landmark. The method of timing should be described in the report, with contact measurements made to an accuracy of 0.5 seconds or better. Observations which are acquired in this manner are particularly important in studies which analyze changes in the Sun's diameter, and for research dealing with the movement of the Moon's nodes.

Although eclipses no longer play the important scientific role that they once did, they will always be magnificent spectacles for both amateur and professional astronomers alike. I hope that each reader will experience one or more for themselves, for only then is the majesty of the star we call Sun at its finest. Perhaps the English poet Alfred Noyes (1880–1958) best expressed the sensation which is engendered by a total solar eclipse, when he wrote:

> In old Cathay,
>   Before the western world began,
> They saw the moving font of day
>   Eclipsed, as by a shadowy fan;
> They stood upon their Chinese wall.
>   They saw his fire to ashes fade,
> And felt the deeper slumber fall
>   On domes of pearl and towers of jade.

References

Abetti, G. 1961, *The Sun*, Macmillan Press, New York, NY.

Akasofu, S. 1982, *Sky & Telescope*, December.

Ammons, R. 1981, *Solar Bulletin*, May.

Astronomical Almanac, US Government Printing Office, Washington, DC.

Babcock, H. W. 1953, *Astrophysical Journal*, **118**, 387.

Babcock, H. W. 1961, *Astrophysical Journal*, **133**, 572.

Bartels, J. 1957, *Annals of the IGY*, **4**, Pergamon Press, London.

Bartels, J. and Chapman, S. 1957, *Nachrichten der Akadem der Wissenschaften in Göttingen II. Mathematisch Physikalische Klasse*, Vandenhoeck und Ruprecht, Göttingen, Federal Republic of Germany.

Batten, A. 1971, *Sky & Telescope*, p. 98, February.

Biermann, J. 1951, *Zentralinst Astrophysik*, **29**, 274.

Bray, R. J. and Loughhead, R. E. 1965, *Sunspots*, John Wiley & Sons Inc., New York, NY.

Carrington, R. 1858, *Monthly Notices of the Royal Astronomical Society*, **19**, 1.

Carrington, R. 1859, *Monthly Notices of the Royal Astronomical Society*, **20**, 13.

Carrington, R. 1863, *Observations of the Spots on the Sun*, p. 221, London.

Chambers, G. F. 1890, *A Handbook of Descriptive and Practical Astronomy*, Clarendon Press, Oxford, England.

Chernan, C. M. 1978, *The Handbook of Solar Flare Monitoring and Propagation Forecasting*, Tab Books, Blue Ridge Summit, PA.

Chernan, C. M. 1980, *73' Magazine*, December.

Chou, B. R. 1981, *Sky & Telescope*, August.

Compton, T. G. 1987, *Journal of the American Association of Variable Star Observers*, **16** (2), 131.

Dellinger, J. H. 1935, *Physical Review*, **48**, 705.

Dialogue Concerning the Two Chief World Systems, Berkeley, CA, (1953 translation of the original work by Galileo).

Drake, S. 1957, *Discoveries and Opinions of Galileo*, Anchor Press, NY.

Eddy, J. A. 1976, *Science*, **192**, 1189.

Eddy, J. A. 1977, *Climatic Change*, No. 1, p. 170.

Eddy, J. A. 1983, *Solar Physics*, **89**, 195.

Eigenson, M. S. 1947, *Priroda*, No. 6, p. 3.

Fishman, G. J. 1988, *Nature*, 4 February.

Gibson, E. G. 1973, *The Quiet Sun*, US Government Printing Office, Washington, DC.

Giovanelli, R. 1984, *Secrets of the Sun*, Cambridge University Press, Cambridge, England.

Gleissberg, W. 1958, *Journal of the British Astronomical Association*, **68**, 148.

Gleissberg, W. 1971, *Solar Physics*, **21**, 240.

Hale, G. E. 1912, *Publications of the Astronomical Society of the Pacific*, **24**, 223.

Hale, G. E. *et al.* 1919, *Astrophysical Journal*, **49**, 153.

Hale, G. E. and Nicholson, S. B. 1938, *Carnegie Institute – Washington Publications*, **49**, 8.

Heckman, G. 1988, *Insight*, Public Affairs Office of Lewis Research Center (NASA), 1 August.

Hewish, A. 1988, *Solar Physics*, **116**, 195.

Hewish, A., Tappin, S. J. and Gapper, G.R., 1985, *Nature*, **314**, 137.

Howard, R. 1968, 'Research on Solar Magnetic Fields from Hale to the Present.' A talk given at Hale Centennial Symposium at meetings of the American Association for the Advancement of Science at Dallas, Texas, December 1968.

Howard, R. and LaBonte, B. J. 1980, *Astrophysical Journal*, **239**.

Huggins, C. H. 1866, *Monthly Notices of the Royal Astronomical Society*, **26**, 260.

Kiepenheuer, K. O. 1953, *The Sun*, ed. Kuiper, University of Chicago Press, Chicago, IL.

Kononovich, E. V., Mironova, I. V. and Serebryakov, B. E. 1986, *Soviet Astronomy Letters*, **12**, 164.

Labitzke, K. 1982, *Geophysical Research Letters*, **14**, 535.

Labitzke, K. and Van Loon, H. 1988, *Journal of Atmospheric and Terrestrial Physics*.

Leighton, R. B. 1964, *Astrophysical Journal*, **140**, 1547.

Leighton, R. B. 1969, *Astrophysical Journal*, **156**, 1.

Liggett, M. A. and Zirin, H. 1985, *Solar Physics*, **97**, 51.

Lincoln, J. V. 1964, *Planetary & Space Science*, **12**, 419.

Maunder, A. S. D. 1940, private communication between A. S. D. Maunder and Stephen A. and Margaret Ionides of Harvard Observatory's Fremont Pass Station, Climax, Colorado (courtesy of Helen Coffey).

Maunder, E. W. 1904, *Monthly Notices of the Royal Astronomical Society*, **64**.

Maunder, E. W. 1922, *Monthly Notices of the Royal Astronomical Society*, **82**, 534.

Mayall, R. N. and Mayall, M. W. 1968, *Skyshooting-Photography for Amateur Astronomers*, Dover Publications, New York, NY.

McIntosh, P. S. 1981, *The Physics of Sunspots*, Sacramento Peak Observatory, Sunspot, NM.

McKinnon, J. A. (ed.) 1987, *The Sunspot Activity in the Years 1610–1987* (revised edition of *The Sunspot Activity in the Years 1610–1960*), World Data Center A, Boulder, CO.

McKinnon, J. A. 1988, *Supplement to The Sunspot Activity in the Years 1610–1987*.

McNish, A. G. and Lincoln, J. V. 1949, *Transactions of the American Geophysical Union*, **30**, 673.

Meeus, J., Grosjean, C. C. and Van Der Leen, W. 1966, *Canon of Solar Eclipses*, Pergamon Press Ltd, Oxford, England.

Menzel, D. H. 1949, *Our Sun*, ed. H. Shapley and B. J. Bok, Country Life Press, Garden City, NY.

Mitchell, S. A. 1923, *Eclipses of the Sun*, Columbia University Press, New York, NY.

Mitra, S. N. 1970, *Solar Physics*, **15**, 249.

Mögel, H. 1930, *Telefunkenztg*, **11**, 14.

Muirden, J. 1975, *Beginner's Guide to Astronomical Telescope Making*, Pelham Books, London, England.

Neidig, D. F. 1983, *Sky & Telescope*, p. 226, March.

Neidig, D. F. 1988, *Solar Bulletin*, September.

Neidig, D. F. and Cliver, E. W. 1983, *A Catalogue of Solar White-Light Flares (1859–1982), Including Their Statistical Properties and Associated Emissions*, (AFGL-TR-83-0257), National Technical Information Service.

Noyes, R. W. 1982, *The Sun, Our Star*, Harvard University Press, Cambridge, MA.

Oppolzer, T. R. 1962, *Canon of Eclipses*, Dover Publications, NY (reprint).

Petrie, W. 1963, *Keoeeit – The Story of the Aurora Borealis*, Pergamon Press Ltd, Oxford, England.

Porter, J. G. 1943, *Journal of the British Astronomical Association*, **53**, No. 2.

Preliminary Report and Forecast of Solar Geophysical Data, 1987, Descriptive Text, 9 June.

Preliminary Report and Forecast of Solar Geophysical Data, 1990, No. 756, 27 February.

Publications of the Astronomical Society of the Pacific, 1947, **59**, 36.

Rao, J. 1988, 'Predictions for the Total Solar Eclipse of 11 July 1991,' PO Box 1122, Linden Hill, NY.

Richardson, R. S. 1948, *Astrophysical Journal*, **107**, 78.

Rosebrugh, D. W. 1951, *Sky & Telescope*, p. 254, August.

Roth, G. D. (ed.) 1975, *Astronomy, A Handbook*, Springer-Verlag Press, West Germany.

Schatten, K. H., Mayr, H. G., Omidvar, K. and Maier, E. 1986, *Astrophysical Journal*, **311**, 460.

Scheiner, C. 1630, *Rosa Ursina sive Sol ex Admirando Facularum*, Apud Andream Phaeum Typographum Ducalem, Bracciani, Italy.

Schove, D. J. 1979, *Solar Physics*, **61**, 423.

Schove, D. J. 1983, *Sunspot Cycles*, Stroudsburg, USA.

Schwabe, H. 1849, *Astronomische Nachrichten*, **21**, 234.

Schwabe, H. 1851, *Bern Mitteilung*, p. 94.

Schwarzschild, M. 1960, *Astrophysical Journal*, **130**, 345.

Shapley, A. H. 1949, *Publications of the Astronomical Society of the Pacific*, **61**, 13.

Sheeley, N. R. and Harvey, J. W. 1981, *Solar Physics*, **70**, 237.

Silverman, S. M. and Mullen, E. G. 1973, *Sky Brightness During Eclipses: A Compendium from the Literature*, Air Force Cambridge Research Laboratory.

Smith, H. J. and Smith, E. v. P. 1963, *Solar Flares*, Macmillan Publishing Co, New York, NY.

Snodgrass, H. and Wilson, P. 1987, *Nature*, **328**, 696.

Solar-Geophysical Data, 1986, Number 499. Explanation of Data Reports, p. 23.

Taylor, P. O. 1985, *Journal of the American Association of Variable Star Observers*, **14**, No. 1.

Taylor, P. O. 1987, *Journal of the American Association of Variable Star Observers*, **16**, No. 2.

Taylor, P. O. 1988a, *Journal of the American Association of Variable Star Observers*, **17**, No. 1.

Taylor, P. O. 1988b, *Journal of the American Association of Variable Star Observers*, **17**, No. 2.

Taylor, P. O. 1989a, *Sky & Telescope*, p. 419, February.

Taylor, P. O. 1989b, *Journal of the American Association of Variable Star Observers*, **18**, No. 1.

Taylor, P. O. 1989c, *Journal of the British Astronomical Association*, October.

Taylor, P. O. 1990, *Journal of the American Association of Variable Star Observers*, **19**, No. 1.

Thiele, T. N. 1859, *Astronomische Nachrichten*, **50**.

Thykier, C. 1988, (Curator) Ole Romers Museum, Copenhagen, Denmark, (personal communication).

Thompson, R. 1989, *Solar Terrestrial Predictions Workshop*, Extended Abstracts, Leura, Australia, S-74.

Todd, D. P. 1899, *Stars and Telescopes*, Little, Brown and Company, Boston, MA.

Van Den Bergh, G., Wilink, T. J. and Zoon, N. V. 1955, *Periodicity and Variations of Solar and Lunar Eclipses*, Haarlem, Netherlands.

155

Vitinskii, Y. I. 1965, *Solar Activity Forecasting*, Israel Program for Scientific Translations, Jerusalem.

Wald, A. 1940, *Annals of Mathematical Statistics*, **11**, 284.

Waldmeier, M. 1947, *Publications of the Zurich Observatory*, no. 9, p. 1.

Waldmeier, M. 1955, *Ergebnisse und Probleme der Sonnenforschung*, 2nd edn, Geest & Portig, Leipzig.

Waldmeier, M. 1961, *The Sunspot Activity in the Years 1610–1960*, Schulthess and Co., Zurich, Switzerland.

Warshaw, D. 1960, *Electronics*, p. 38, June.

Webb, T. W. 1893, *Celestial Objects for Common Telescopes*, 5th edn., **1**, Longmans, Green & Co, London.

Williams, D. G. 1985, *Australian Journal of Physics*, **38**, 1027.

Willson, R. and Hudson, H. 1988, *Nature*, 28 April.

Wilson, R. M. 1987, *Solar Physics*, **111**, 255.

Wolf, A. R. 1852, *Naturf. Gesell. Bern Mitt.*

Wolf, A. R. 1858, *Astronomische Mitteilungen Eidgenössisch Sternwarte*, No. 10, p. 6.

Young, C. A. 1888, *The Sun*, D. Appleton and Company.

Zirin, H. 1972, *Solar Physics*, **22**, p. 34.

Zirin, H. 1988, *Astrophysics of the Sun*, Cambridge University Press, Cambridge, England.

Zirker, J. B. 1984, *Total Eclipses of the Sun*, Van Nostrand Reinhold Company, New York, NY.

Index